GEOSYNTHETICS IN CIVIL ENGINEERING

GEOSYNTHETICS IN CIVIL ENGINEERING

Translated by
GERARD P.T. M. VAN SANTVOORT
Van Santvoort Consultancy B.V., Rosmalen, Netherlands

A.A. BALKEMA / ROTTERDAM / BROOKFIELD /1995

CUR/NGO Report151

Centre for Civil Engineering Research and Codes
Netherlands Geotextile Organization

The pictures in this book are made available by AKZO-NOBEL, CeCo, CTN Netherlands, Intercodam, Nicolon, Pelt & Hooykaas, Ministry of transport, Public Works and Watermanagement, Rook and Van Santvoort Consultancy.

Every effort has been made to ensure that the statements made in this publication provide a safe and accurate guide, however no liability of any kind can be accepted in this respect by the publisher or the author.

Published by
A.A. Balkema, P.O. Box 1675, 3000 BR Rotterdam, Netherlands (Fax: +31.10.4135947)
A.A. Balkema Publishers, Old Post Road, Brookfield, VT 05036, USA (Fax: 802.276.3837)

ISBN 90 5410 604 2

Table of contents

PART 3: ELABORATED EXAMPLES

Preface

The civil engineering branch is still little acquainted with the possibilities of geosynthetics. This was the reason for the Centre for Civil Engineering Research and Codes (CUR) in the Netherlands and the Geosynthetic Organization (NGO) to edit a handbook about this subject. The first edition appeared in 1991 in the Dutch language and is in use at many Technical Colleges. This handbook has been achieved under the responsibility of a mixed CUR/NGO Commission 'C 57 Handbook Geotextiles and Geomembranes'.

This publication has been translated by G. P.T. M. van Santvoort. By means of this English version, a much larger group of students and civil engineering consultants can be acquainted with the application of geotextiles.

July 1995
The Board of the CUR
The Board of the NGO

List of symbols

B	Width in m
C_u	Undrained shear strength, kN/m^2
E	Modulus of elasticity, N/mm^2
F	Force per unit of width, kN/m
F_a	Anchoring force, kN
G	Loading, kN/m^2
H	Height or thickness, m
L	Distance , m
L_a	Anchoring length, m
M	Moment, kNm
$O(p)$	Characteristic pore dimension, mm or μm
P_s	Loss of pressure height at standard temperature and standard filter velocity, m
T_s	Standard temperature, 10°C
T	Wave period, s
W	Friction number, %
a	Lever arm of force F, m
d	Grain-size or -diameter, mm or μm
d_p	Sieve number, mm or μm
e	Thickness of a geosynthetic, mm
f	Friction coefficient
f_s	Safety factor or -coefficient
g	Acceleration of gravitation, m/s^2
Δh	Hydraulic head difference, m or mm
i	Hydraulic gradient
k	Coefficient of permeability, m/s
p	Rest on sieve, %
q_c	Vertical cone value, MN/m^2
r_i	Reduction factor
r	Radius of slip-circle, m
v	Current velocity

α   Slope angle, °
δ   Angle of friction between geosynthetic and surrounding soil, °
ρ   Mass, kg/m^3
σ   Tension, kN/m^2
φ   Angle of shearing resistance, ° (internal friction)
ψ   Permittivity, s^{-1}

CHAPTER 1

Introduction

1.1 *Objective and target group*

In the past 25 years many applications of geosynthetics have proved their value in civil engineering projects. To meet the demand for knowledge concerning geosynthetics, a comprehensive manual has been published in 1986 called 'Geotextiles and Geomembranes in Civil Engineering'. This book was revised in early 1994.

For students at Technical Colleges there appeared to be a demand for an easy to read and to consult, concise handbook. This handbook offers students in civil engineering, environmental technics and land development the subject matter necessary for an orientation on the application possibilities of geosynthetics as building material. The presented general knowledge of geosynthetics may also be of value for officials from Authorities, Consulting Engineering firms, Contractors and the Supply Industry concerned. All those who like to widen their knowledge in theoretical aspects are referred to the above mentioned comprehensive manual and to the publications mentioned in the literature list of this book.

This book deals with the application of geosynthetics in soil structures, foundations engineering and bank- and bed protection. The application of geosynthetics for other civil engineering purposes is mentioned only obliquely, like asphalt reinforcement and drainage of embankments.

Besides, it has to be mentioned that the application possibilities of geosynthetics is still developing. In this book the state of the art of the mid-nineties is presented.

1.2 *Framework of the book*

It is assumed that students command the basic knowledge of geotechnics and hydraulics. The book is in three parts:

The first part 'General', with the Chapters 2 and 3, forms a complete whole. It contains the elementary knowledge of geosynthetics every civil student should know.

Part two 'Design Considerations' is an elaboration of Part one and is concentrated on applications of geosynthetics in civil engineering practice. The lecturer may make a choice out of the Chapters 4, 5 and 6, in which the three most important functional applications are presented, respectively: reinforcement, filter/separation and screen.

Each chapter starts with the characteristic material properties which are of functional importance for the applications discussed. Thereafter design considerations are presented, followed by a section concerning remaining aspects. Because these aspects refer to functional applications, they are presented in Chapters 7 and 8 again in a well-ordered and comprehensive way.

In part three, 'Elaborated Examples', three structural designs are elaborated in detail with calculation examples. The Chapters 9, 10 and 11 are related to, respectively, the Chapters 4, 5 and 6.

1. General

CHAPTER 2

Examples

2.1 *Introduction*

For a quick orientation and introduction on application possibilities of geosynthetics eight typical examples are presented in the following section. The first three examples concern soil reinforcement, the next two a filter or separation layer and the last three an impermeable screen.

With each example a picture and a figure of an application is shown and the main functions of the geosynthetic are mentioned. For the meaning of the given numbers one is referred to the corresponding, larger drawing given with the in Section 3.4 elaborated examples.

2.2 *Examples*

Example 1: Reinforcement 1.

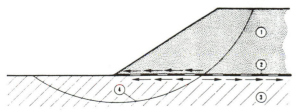

Figure 2.1. Embankment on a natural ground with low bearing capacity; polyester fabric.

Example 2: Reinforcement 2.

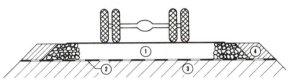

Figure 2.2. Road foundation on a natural ground with low bearing capacity; polyester fabric.

Example 3: Reinforcement 3.

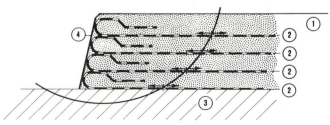

Figure 2.3. Steep slope structure; polyester fabric.

Example 4: Filter 1.

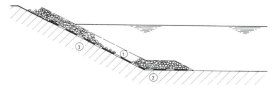

Figure 2.4. River bank protection with a filter construction and a riprap revetment; polypropylene, polyethylene or polyester fabric or nonwoven.

Example 5: Filter 2.

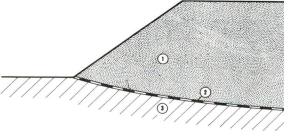

Figure 2.5. Separation of embankment and natural ground; polyester, polyethylene or polypropylene nonwoven.

Example 6: Screen 1.

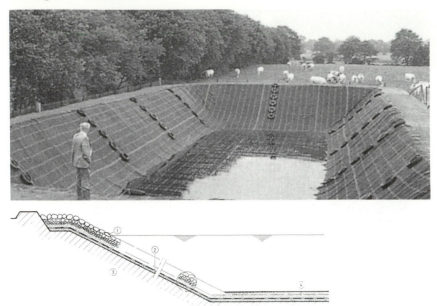

Figure 2.6. Sealing of a reservoir; polyethylene membrane.

Example 7: Screen 2.

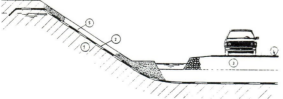

Figure 2.7. Road in cutting below water table; polyethylene or polyvinylchloride membrane.

Example 8: Screen 3.

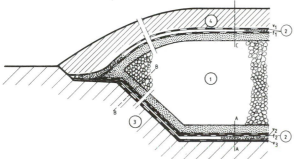

Figure 2.8. Landfill; polyethylene membrane; protecting nonwovens of polyester or poly-propylene.

Examples of other applications used in construction and civil engineering
Geosynthetics are used for many purposes and in a large variety of forms which
are not the subject of this book. Some examples are:

Vertical drains:
– For acceleration of the consolidation and consequently the settlement of
  embankments.
– For the run-off of water along structures.

Horizontal drains:
– For the acceleration of the discharge of free water in soil structures.

Inlays in road pavements:
– To decelerate reflective cracking.
– As reinforcement of an asphaltic layer.
– As carrier for a wearing course.

Temporary boardings:
– To let cure cement-bound materials in a certain shape.

Containers:
– Gabions units for hydraulic structures filled with granular material.

Honeycomb structures:
– Open cellular structures filled with granular material, creating rigid revetments to prevent erosion on banks.
Pictures of some of these applications have been placed all through the book.

2.3 *Summary*

Eight characteristic examples are presented of applications of geosynthetics in civil engineering practice. The examples are functionally clustered.

Reinforcement:
– Embankment on a natural ground with low bearing capacity.
– Road foundation on a natural ground with low bearing capacity.
– Steep slope structure.

Filter:
– Bank protection with a filter construction.
– Separation of embankment and natural ground.

Screen:
– Sealing of a reservoir.
– Road in cutting below water table.
– Landfill.

Other functional applications are not a subject of this book.

CHAPTER 3

Properties and functional applications

3.1 Introduction

Geosynthetics are used for several purposes in civil engineering, especially as reinforcement, as filter or separation layer, or as a screen. Dependent on the functions to be performed, different requirements are requested of the properties of geosynthetics.

In Section 3.2 a review is given of several basic materials, the production process and end products.

In Section 3.3 an explanation is given of the functional applications and the properties of end products.

In Section 3.4 the examples, presented in Chapter 2 are elaborated.

In Section 3.5 the potential changes in properties of geosynthetics are discussed, particularly the changes by alteration, creep, hydrolysis, mechanical damage and chemical and biological attack.

In Section 3.6 the remaining aspects are discussed, particularly splitting of functions, dimensions, seams, environmental factors and quality control.

In Section 3.7 a summary is given.

3.2 Basic materials, processing and end products

Basic materials
There are five main polymers used in the manufacturing of geosynthetics:

Polyester	(PET)
Polypropylene	(PP)
Polyethylene	(PE), with the species HDPE ('high density') and LDPE ('low density')
Polyamide*	(PA), with the species PA6 and PA 6.6
Polyvinylchloride	(PVC)

*Products of polyamide are scarcely applied nowadays.

11

Table 3.1. Some properties of basic materials.

Basic material	Unit mass kg/m^3	Tensile strength at 20°C; N/mm^2	Modulus of elasticity N/mm^2	Strain at break in %
PET	1380	800-1200	12000-18000	8-15
PP	900	400-600	2000-5000	10-40
PE LDPE	920	80-250	200-1200	20-80
HDPE	950	350-600	600-6000	10-45
PA	1140	700-900	3000-4000	15-30
PVC	1250	20-50	10-100	50-150

The basic materials consist mainly of the elements carbon, hydrogen and sometimes nitrogen and chlorine (PVC); they are produced from coal and oil.

Some characteristic properties are presented in Table 3.1.

Production process

The polymers come from a chemical plant in the shape of granules. During further processing the granules are melted and after extrusion, eventually followed by spinning, the semi-manufactured products get the shape of sheets, tapes (broad or very small) or threads.

During melting before extrusion and during the further processing additives can be used. These additives have the purpose of improving the less favourable properties of the basic materials, especially the sensitivity for alteration (see Section 3.5). Pigments can also be added. In the following phases the semi-manufactured products are made: Threads (mono-filaments), yarns, fibres, tapes and membranes. The final processing consists of weaving, stitching to nonwovens, and if required the welding of membranes into a larger width. A simplified processing scheme is presented below.

Next to the most important end products woven fabrics and nonwovens, grids, mattings and composites have to be mentioned.

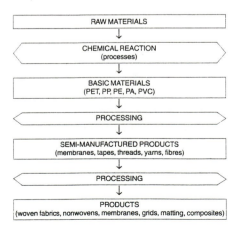

End products

Woven fabrics are produced either from tapes (tape-fabric) or yarns. A fabric consists of wrap yarns and weft yarns. Depending on the end-use threads of several polymers can be used in a fabric, so that specific properties of the different polymers can be combined. Woven fabrics are water-permeable and soil-tight or not soiltight, depending on the dimensions of the mesh in combination with the grain-size distribution of the soil.

Nonwovens are textiles produced by mechanical, chemical or thermal bonding or interlocking of short fibres (sometimes long fibres). The mechanical interlocking involves the use of a large number of barbed needles (needle punching), which are pressed up and down through a package of fibres so that the fibres are strongly intermingled. Nonwovens can also be chemically bonded. In this system a binder is provided which hardens when subjected to high temperatures. Also thermally bonded nonwovens are made by heating fibre mixtures of different melting temperatures (different polymers) under pressure. Combinations of the above mentioned technics are also applied. Nonwovens are water-permeable and soil-tight.

Woven fabrics and nonwovens together are called '*geotextiles*'.

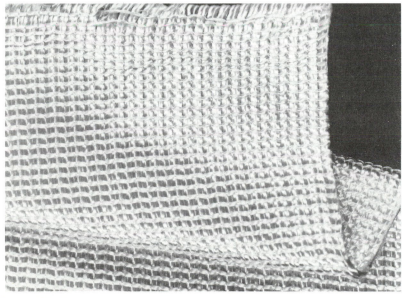

Woven fabric

Woven fabric of tapes

Membranes are thin two-dimensional sheets with a very low permeability, called '*geomembranes*'.

At several places in this book pictures are presented of constructions in which geotextiles and geomembranes are applied.

Grids are coarse-meshed lattices of parallel-run yarns thermically or mechanically (by weaving) connected at the crossings. Thick, extruded sheets, too, are perforated in a matrix system and stretched. Grids are very permeable and not soil-tight. They are an alternative for woven fabrics if the soil-tightness is of no importance. Two-sides stretched grid (biaxial lattice) is amongst other applications, used for the manufacturing of gabions.

Mattings are three-dimensional mats produced by extruding monofilaments into a rotating profile roller, followed by cooling. As a result the yarns stick together at crossings which are spatially arranged. Mattings are very permeable but they can prevent the wash-out of soil particles (erosion control).

Composites appear in many shapes and species. The objective is to combine the favourable properties of a single geosynthetic with other materials. Also woven fabrics with geomembranes are combined to form strong, watertight composites (see drawing on page 27). Mattings can be combined with geomembranes resulting in soiltight, water-transporting composites. Steel wire can enlarge the tensile strength and the shape stability when combined with a woven fabric.

An overview of end products is presented in the Product Catalogue of the NGO [2].

Grid

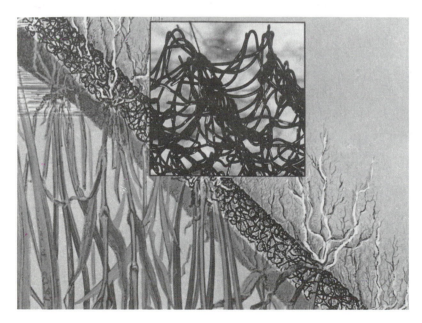

Matting

3.3 *Functional applications and properties of end products*

When designing civil engineering constructions, the functions to be performed have to be analysed first; after that the suitable materials and products can be selected. When geosynthetics are provided the properties like strength and elongation are derived from the basic materials (polymers) and from the shape of the product (permeability and soil-tightness).

For instance, the reinforcement of a soil structure requests a strong, relatively stiff and preferably water-permeable material. Then a woven fabric of polyester is a logical choice. For a filter/separation function the material has to be flexible, water-permeable and soil-tight. A nonwoven or a light-weight woven fabric of polyethylene is the material to be chosen.

A screen function requests a waterimpermeable geosynthetic like a geo-membrane of polyethylene.

It has to be noticed that in most applications a geosynthetic fulfils a main function and a minor function. For instance, a geosynthetic with a filter function has to absorb tensile stresses; see Section 3.6.

The relation between properties and functional applications of different geo-synthetics are presented in Table 3.2.

Table 3.2. Relation between properties and funcional applications.

Product	Properties	Reinforcement	Filter	Screen
Woven fabric	Strength, stiffness Water-permeable Soil-retaining	x	x	–
Nonwoven	Ductile/elastic Soil-retaining Water-permeable	–	x	–
Geomembrane	Ductile/elastic Soil-tight Water-impermeable	–	–	x
Basic materials				
PET		x	x	x
PP		–	x	x
PE		–	x	x
PA		–	x	x
PVC		–	–	x

3.4 *Elaboration of the examples*

The eight examples of functional applications presented in Chapter 2 are elaborated below.

REINFORCEMENT

All geosynthetics applied as reinforcement have the main task of preventing the collapse or inadmissible deformation of soil structures. These geosynthetics are applied mainly for embankments on natural grounds with low bearing capacity and for the construction of steep slopes. The main properties of geosynthetics for reinforcement are: Strength, stiffness and the conservation of these properties during the life cycle of the construction. Water-permeability is often a minor demand. Woven fabrics of polyester meet the requirements very well. Depending on the kind of soil and the particular application also grids can be applied.

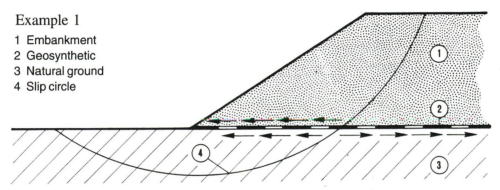

Example 1
1 Embankment
2 Geosynthetic
3 Natural ground
4 Slip circle

Figure 3.1. Embankment on a natural ground with low bearing capacity.

Objective:
– Prevention of embankment-edge failure caused by the collapse of the subsoil.
Effect:
– After an initial deformation the woven fabric delivers a reaction force as a contribution to the resistance against the slide-down of the embankment.
Forces:
– Tensile strain in the fabric on the spot of the crossing with the slip circle; the tensile strain has to be transmitted via friction to the soil at both sides of the fabric; from this the requested anchoring length can be calculated.
Particular details:
– Consolidation of the subsoil can make the reinforcement superfluous in due time.

Requirements:
– Large tensile strength in combination with little elongation and creep.
– Large friction between the soil of the embankment and the subsoil.
– Sufficiently water-permeable to make possible the consolidation of the subsoil and the reduction of the water-pressure.

Material:
– A woven fabric of polyester with a high modulus of elasticity.

Execution:
– The fabric is unrolled with a overlap in the direction at right angles to the future slope.

Example 2

1 Sub-base of coarse granular material
2 Geosynthetic
3 Subsoil
4 Shoulder

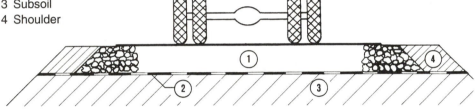

Figure 3.2. Road foundation on a subsoil with low bearing capacity.

Objective:
– Improvement of the bearing capacity of the road and separation of coarse sub-base material and subsoil.

Effect:
– After an initial deformation the geosynthetic delivers a tensile force contributing to the load distribution on the subsoil.
– The geosynthetic prevents the mixing of subsoil particles with material in the sub-base.

Force:
– Tensile strain in the geosynthetic; the transmission of the tensile strain in the geosynthetic to the surrounding material is realised by the friction between the geosynthetic and the soil at both sides of the geosynthetic.

Particular details:
– The duration of the reinforcement depends on the local soil condition and the function of the road. The effect of the separation function is more durable than the effect of reinforcement.

Requirements:
– Large tensile strength and low creep behaviour.
– Low creep in case of a permanent reinforcement function.
– Large friction.
– Sufficiently water-permeable to permit the consolidation of the subsoil.

Material:
– A polyester woven fabric with a high modulus of elasticity; in case the geosynthetic is in between layers of coarse material also a grid may be applied.

Execution:
– Overlapping slips with a width of circa 5 meters are placed on the sub-base at right angles to the future road direction.

Example 3
1 Embankment
2 Geosynthetic
3 Subsoil
4 Revetment

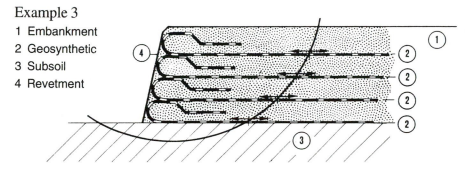

Figure 3.3. Construction of a steep slope.

Objective:
– Prevention of embankment-edge failure.

Effect:
– After an initial deformation the geosynthetic delivers the reaction force against the slide-down of the slope by locking the soil in small compartments and anchoring in the embankment behind.

Forces:
– Tensile strain in the geosynthetic on the spot of the crossing of the geosynthetic with the slip-circle; the tensile strain is transmitted to soil by friction between the geosynthetic and the surrounding soil; hence the necessary anchoring length can be calculated.

Particular details:
– Because of the permanent reinforcement function high demands are made upon the durability of the geosynthetic.
– Measures have to be taken against wilful damage.
– The front side has to be protected against UV-radiation by a revetment.

Requirements:
– Large tensile strength, little elongation and creep.
– Large friction.
– Water-permeable.
– Soil-tight.

Material:
– A polyester woven fabric with a high modulus of elasticity. If coarse material is applied in the embankment also a grid can be used.

Execution:
– Slips are placed at right angles to the embankment direction. The slips are placed with an overlap or sewed.

FILTER AND SEPARATION

Geosynthetics are applied as filter and separation layer in the following circumstances:
– For bank and bed protection.
– Erosion prevention on slopes.
– As a soil-tight layer behind a pile-planking.
– To separate soil-layers in an embankment.

The main properties to fulfil the filter and separation functions are the soil-tightness and the water-permeability.

One speaks of a filter function when the main function of the geosynthetic is to retain soil-particles and to let pass the water, for instance application of a geosynthetic behind a pile-planking. A function is called separation function when the geosynthetic has to separate two different kinds of soils, for instance in embankments.

Nonwovens and light-weight woven fabrics are the most suitable products for filter and separation purposes. Polypropylene, polyethylene, polyester and polyamide are proper materials.

Example 4

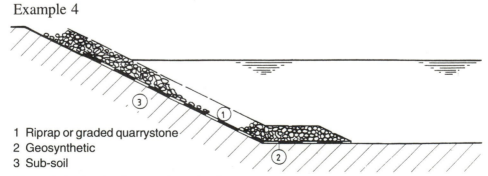

1 Riprap or graded quarrystone
2 Geosynthetic
3 Sub-soil

Figure 3.4. Riverbank protection with filter construction.

Objective:
- Prevention of erosion of bank and bottom.

Effect:
- The geosynthetic retains soil particles and is water-permeable.

Forces:
- Leading tensile forces occur during execution.

Particular details:
- Damage of the geosynthetic during dumping of stones has to be prevented.

Requirements:
- Soil-tight.
- Water-permeable.
- Sufficiently elastic to follow the roughness of the subsoil.

Material:
- A nonwoven or woven fabric of polypropylene, polyethylene or polyester.

Execution:
- Slips sewed together to widths of 20 m are unrolled at a right angle to the slope; the submerged part is sunk like a brushwood mattress.

Example 5

1 Embankment
2 Geosynthetic
3 Natural ground

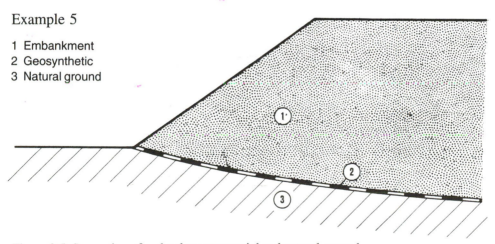

Figure 3.5. Separation of embankment material and natural ground.

Objective:
- Prevention of mixing of subsoil with material of the embankment.

Effect:
- The geosynthetic separates the soil layers and is water-permeable.

Forces:
- Leading forces occur in the geosynthetic during execution.

Particular details:
- The durability of the separation function depends on the method of execution and the constitution of the natural ground.

Requirements:
– Soil-tight.
– Water-permeable.
– Sufficiently elastic to be able to follow the settlement of the natural ground.
Material:
– Nonwovens of polyester, polyethylene or polypropylene.
Execution:
– Overlapping slips are unrolled longitudinally to the embankment.

SCREEN

Geosynthetics are applied as water-impermeable layers for:
   – Revetments of reservoirs to prevent leakage.
   – Enwrapping of trenched constructions to prevent the inflow of water.
   – Isolation of landfills and polluted areas to prevent the outflow of polluted water and the inflow of rainwater.
   The main property for a geosynthetic applied as a screen is the water(liquid) tightness. Geomembranes are the products for these applications.
   Because geomembranes have to fulfil their function during a long period of time and sometimes in an aggressive environment, special provisions have to be taken to prevent leakage caused by poor execution and sudden settlement and deformation of the natural ground. These provisions can be provided by reinforcement of the geomembrane with a strong woven fabric and covering of the membrane by a sand-layer to prevent mechanical damage. Sometimes several membranes are applied with a drainage interlayer.

Example 6

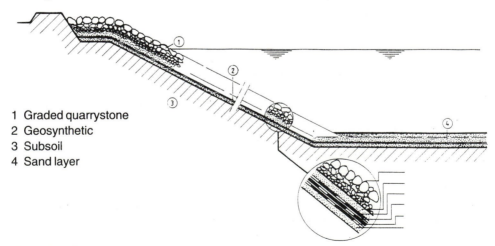

1 Graded quarrystone
2 Geosynthetic
3 Subsoil
4 Sand layer

Figure 3.6. Revetment construction of a water reservoir.

Objective:
– Prevention of leakage (for instance reservoirs for drinking water or irrigation water) and of inflow of polluted water from the surrounding ground water.

Effect:
– The geosynthetic prevents the transport of water to the surroundings.

Forces:
– In case of unequal support tensile forces occur in the geomembrane caused by differences in water pressure.
– Depending on the method of execution tensile forces may occur during application.

Particular details:
– The geomembrane has to be protected against mechanical damage during execution as well as during the lifetime of the construction and against root-grow.

Requirements:
– Water-impermeable.
– Elastic.
– Durable.

Material:
– A geomembrane of polyethylene.

Execution:
– Geomembrane slips are welded together in situ.

Example 7

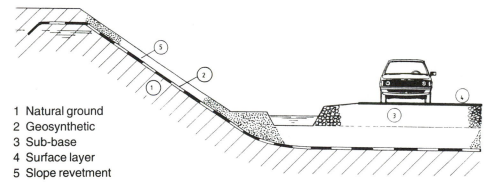

1 Natural ground
2 Geosynthetic
3 Sub-base
4 Surface layer
5 Slope revetment

Figure 3.7. Road construction in a trench below ground water table.

Objective:
– Prevention of the inflow of ground water in the trench and maintaining of the water table outside the trench at the original level.

Effect:
– The geosynthetic prevents the water inflow into the trench.

Forces:
- Tensile strain in the geomembrane during application depending on the method of execution.

Particular details:
- Due to the water pressure caused by the groundwater the geomembrane has to be ballasted to prevent the floating of the membrane.

Requirements:
- Water-impermeable.
- Elastic.
- Durable.

Material:
- A geomembrane of polyethylene or polyvinylchloride.

Execution:
- Slips of the geomembrane are welded together in situ.

Example 8

1 Refuse
2 Geosynthetic
3 Natural ground
4 Sand layer

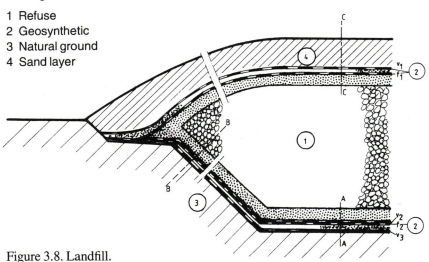

Figure 3.8. Landfill.

Objective:
- Prevention of emission of polluted liquids.
- Geomembrane f1 prevents the inflow of rainwater.
- Geomembrane f2 prevents the emission of pollution to the subsoil and the groundwater.

Effect:
- The geosynthetic prevents the transport of liquids to the surroundings.

Forces:
- Tensile strain in the geomembrane depends on the method of execution.

Particular details:
– The function of geomembrane during a long time is safeguarded by the application of a drainage layer between nonwoven v2 and geomembrane f2 and upon nonwoven v3. When geomembrane f2 leaks unexpectedly, the leaching water can be drained by pumping. Besides, the drainage layer provides an inspection possibility.

Requirements:
– Impermeable.
– Chemical resistant.
– Durable.
– Elastic.

Material:
– A geomembrane of polyethylene.

Execution:
– Slips of membranes are welded in the factory as long and as wide as possible to reduce welding in situ.

3.5 *Transformation of properties*

This section discusses the transformation of properties which are important for the application of geosynthetics.

The properties of geosynthetics can change disfavourably, partly in course of time, partly at a certain moment, by means of alteration, creep, hydrolysis, mechanical damage and chemical and biological attack. These phenomena have to be taken into account when geosynthetics are selected. For instance, a reduction factor has to be introduced in the calculation for the decline of strength caused by these phenomena. This reduction factor does not replace the soil mechanical safety factor (see Section 4.2).

Aging
Raised temperature and ultraviolet radiation have a negative effect on geosynthetics because they stimulate oxidation by which the molecular chains are cut off. If this process has started the molecular chains degrade continuously and the original molecular structure changes. It involves a substantial reduction of the mechanical resistance. The geosynthetic becomes brittle. This phenomenon is called 'aging'. Some basic materials are more sensitive to aging than others; see Table 7.1. To reduce the sensitivity for aging anti-oxidising agents and UV-stabilizers are added during the production process. A well known anti-oxidising agent is carbon black. Some stabilizers, including carbon black, have a negative effect on the mechanical properties of a geosynthetic. In each phase of the production process of a geosynthetic the temperature is raised for the processing of basic materials and half-products. It may result in the start of the aging process. Therefore, quality control has to be performed on the end products.

Based on standard testing methods geosynthetics can be compared with each other, but it is difficult to use the results for a realistic life-cycle calculation. Hence only a qualitative comparison of the most used basic materials is presented in Table 7.1.

The following aspects have to be taken into account when geosynthetics are applied:
– The temperature which may appear during application and the time of exposure;
 – Exposure to sunlight, the duration and intensity;
 – The possibility of leaching of anti-oxidising agents and UV-stabilizers, resulting in subsoil pollution;
 – The possibility of the presence of metals in the surroundings of the geosynthetic, which can act as catalysts in an aging process.

It is not possible to calculate and to express in a figure the decay of properties of a geosynthetic (for instance the strength). During the design period these phenomena have to be considered and taken into account when selecting the type of the basic material.

Creep

Creep is the increase of length of a material under permanent loading. Two important factors are to be mentioned:
 – The material can stretch under permanent loading to such an extent that the material breaks or tears;
 – Under the same circumstances in the course of time one material stretches more than another.

In fact the creep characteristics of a material determine the applicability and the maximum admissible stress of a geosynthetic used as reinforcement. This matter is discussed in Section 4.2.

Hydrolysis

Some geosynthetics like nylon (polyamide) and, to a less extent, polyester are sensitive to hydrolysis under wet conditions (reaction with water). At moderate temperatures a loss of strength of 5% has to be calculated. A rapid decline in strength occurs at temperatures above 80°C.

Mechanical damage

During application of a geosynthetic there is a great chance for damage, which may result in loss of strength or leakage. Hence the instructions have to cover the possibilities for careful execution. Also storage and transport ask for special attention. For instance the dumping of stones at a riverbank revetment brings with it the danger for punching, while caterpillar driven traffic may lead to the tearing of the geosynthetic. In Sections 4.8, 5.5 and 6.4 some particular damage phenomena are mentioned for reinforcement, filter and screen; in Section 7.2 the mechanical damage is pursued in general terms.

Chemical and biological attack

Geosynthetics have to be resistant to the chemicals and micro-organisms present in the surrounding soil. Under some conditions the strength of a woven fabric and the water-tightness of a geomembrane can be affected substantially.

For instance, reinforcing materials of polyester are strongly attacked under high-alkaline conditions. Polypropylene can be attacked by some fungi in such a way that threads, fibres or membranes split (fibrillation). Table 7.1 presents a general view of the resistance of geosynthetics against affection.

3.6 *Remaining aspects*

In addition to functional properties, mentioned in Section 3.3, and the change in properties mentioned in Section 3.5, some other limiting conditions are of importance for the choice of a geosynthetic and the method of instalment. In this paragraph the most important limiting conditions are discussed.

Separation of functions

Usually a geosynthetic is chosen on the basis of requirements for a main function, for instance a nonwoven for a filterfunction. But it may occur that the geosynthetic chosen to meet the requirements of the main function does not match certain secondary loadings, for instance mechanical loading. Sometimes the geosynthetic cannot meet the requirements of a secondary function like the bearing of tensile strain. In such a case the so-called 'separation of functions' can be necessary. This may lead to the choice of another type of geosynthetic (for instance a composite), which meets the requirements for both functions. Another possibility is the application of two different types of geosynthetics which together can perform the required functions.

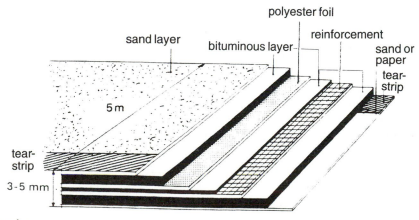

Composite

Unrolling of a geomembrane with an overlap.

Finite dimensions
Geosynthetics are finite and therefore it is necessary to make connections or overlaps. Seams and overlaps are weak spots in a construction and vulnerable. Therefore, they have to be limited as much as possible. If geosynthetics are applied as reinforcement, seams at right angles to the direction of the leading force are unacceptable. In Sections 4.8, 5.5 and 6.4 this matter will be discussed in detail. Also the termination of a geosynthetic and the connection to another part of the construction asks for special attention.

Damage to the environment
Geosynthetics may burden the environment. On the one hand, with a careless instalment remaining parts may get scattered into the surrounding so that, for instance, animals get entangled or ship propellers jammed. On the other hand, toxic additives like softeners, anti-oxidants, UV-stabilizers and compounds with chloride may leach and pollute the environment. In Section 7.4 is discussed the cautiousness which has to be practized during removing, re-use or incineration of geosynthetics. Separate collection prevents that geosynthetics are regarded as chemical waste.

Quality control
To verify whether the geosynthetics meet the prescribed requirements, quality control has to be performed. In most cases the presentation of a certificate is sufficient. In Chapter 8 this matter is elaborated.

3.7 *Summary*

Woven fabrics, nonwovens, geomembranes, grids, mattings and composites can be produced from several types of basic materials. Each end product has at least one main functional property; strength, soil-tightness, permeability or impermeability. For applications in civil engineering projects the most important functions are reinforcement, filtering or separation and screen.

Woven fabrics can perform reinforcing functions as well as filter functions. Nonwovens have a filter and separation function. Geomembranes are able to form an impermeable screen. Sometimes it becomes evident that geosynthetics, chosen on the basis of a main function, do not meet the requirements for a secondary function. In this case one can choose for another geosynthetic which does fulfil the requirements for the main and secondary function or one can choose several geosynthetics which, each in part and together, meet all requirements. The last choice is called 'separation of functions'.

In the course of time the properties of geosynthetics may change under the influence of aging, creep, hydrolysis, mechanical damage and chemical and biological attack.

Geosynthetics are finite. Seams, connections and overlaps are weak spots and form a limitation to the applicability of the product.

Negligent execution, scattering of remains and leaching of toxic elements from geosynthetics may have a harmful influence on the environment.

2: Design considerations

CHAPTER 4

Reinforcement function

4.1 *Introduction*

In Section 3.4 three examples of the reinforcement function of geosynthetics were presented:
- Embankment on a natural ground with low bearing capacity;
- Road foundation on a subsoil with low bearing capacity;
- The construction of a steep wall.

The objective of soil reinforcement is to prevent landslide or unacceptable deformation of the soil structure. This chapter deals with the possibilities of geosynthetics to prevent this.

In Section 4.2 those functional properties are discussed that have to be fulfilled by geosynthetics to perform the reinforcement function. Besides tensile strength and modulus of elasticity (stiffness) the influence of creep and aging on geosynthetic properties are discussed.

Section 4.3 deals with the performance of the reinforcement function of a geosynthetic and how the stresses present in the geosynthetic can be transmitted to the surrounding ground (anchoring of the geosynthetic).

In Section 4.4 a number of design criteria are discussed taking into account the analyses of the acting forces and the several time dependent factors.

In Sections 4.5, 4.6 and 4.7 design calculations are presented which relate to the three examples of reinforcement function.

Section 4.8 deals with the remaining aspects like seams, overlaps and damage. A summary is given in Section 4.9.

4.2 *Functional properties*

Large strength and stiffness are the required properties for the geosynthetic to be able to function as reinforcement. The strength and stiffness have to be maintained during the whole period the geosynthetic has to function as reinforcement. Three factors appear to be of importance: Kind of basic material, temperature and creep. From the table and diagram of Section 4.1 can be read that polyester is the basic material with the largest strength and the highest stiffness.

Applying a reinforcing fabric in a marshland for an embankment (Indonesia)

Table 4.1. Tensile strength, modulus of elasticity (E-modulus), and creep of basic materials at 20°C.

basic ma-terial	tensile strength in N/mm^2	E-modulus in N/mm^2	Admissible percentage of tensile strain due to creep after		
			2 years	10 years	100 years
PET	800-1200	12000-18000	70	65	50
PP	400-600	2000-5000	<20	0	0
PE	80-600	200-6000	<20	0	0
PA	700-900	3000-4000	60	55	40
PVC	20-50	10-100	0	0	0

Compare: The tensile strength and E-modulus of construction steel are, respectively, 500 and 210 000 N/mm^2, of high strength steel, respectively, 1700 and 210 000 N/mm^2.

The actual temperature has a great influence on the strength properties of geosynthetics, which is shown in Sections 4.3 and 4.4. In these diagrams it is made clear that the allowable strain in a construction with geosynthetics has to be much lower than the breaking strain of geosynthetics due to creep behaviour. Phenomena like aging, chemical attack, mechanical damage and hydrolysis also lead to a reduction of the allowable strain in the geosynthetic.

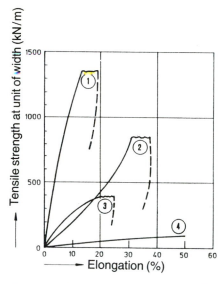

1 Polyester woven fabric
2 Polyamide woven fabric
3 Polypropylene woven fabric
4 All nonwovens

Figure 4.1. Load-elongation curves of common applied woven fabrics and nonwovens.

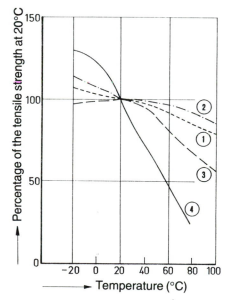

1 Polyester
2 Polyamide
3 Polypropylene
4 Polyethylene

Figure 4.2. Influence of the temperature on the tensile strength.

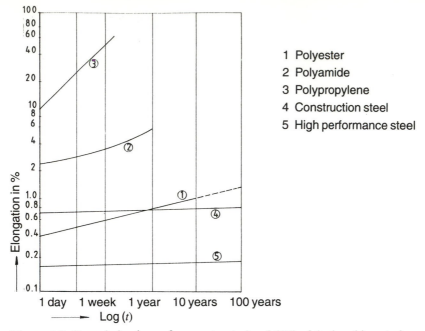

Figure 4.3. Creep behaviour of yarns at a strain of 60% of the breaking strain.

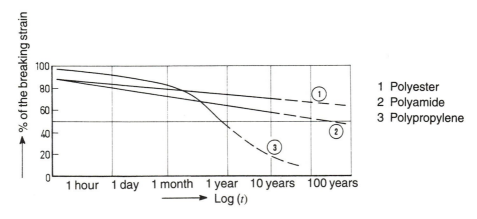

Figure 4.4. Creep behaviour of yarns; breaking strain as function of time.

From Figure 4.4 can be read that polypropylene and polyethylene collapse due to their creep behaviour at very short notice. It may be clear that polyester is the proper material for reinforcement.

Table 4.2. Reduction factors for the determination of the requested tensile strength of a polyester geosynthetic applied as reinforcement.

Kind of loading	Reduction of tensile strength in %	Admissible strain in % of breaking strain	Reduction factor r_i
Creep after 100 years	50	50	$r_k = 2$
Mechanical and chemical	10-90	90-10	$r_m = 1.1\text{-}10.0$
Hydrolysis	5	95	$r_h = 1.05$
Seam	25-50	75-50	$r_n = 1.33\text{-}2.0$

The in Section 3.5 defined reduction factor r_i for creep with an expected life-time for a construction of 100 years for polyester is 2. The decay in strength due to mechanical damage and chemical attack depends on storage, transport and execution circumstances and may vary in practice from 10% to 90% of the tensile strength. When treated with care a reduction factor of 1.1 is allowable. Hydrolysis may lead to a reduction in tensile strength of 5% (see Section 3.5). If a seam occurs in the direction of the strain a reduction in tensile strength of 25% has to be accounted for [2]. In Table 4.2 the above mentioned figures are recapitulated.

For a permanent construction under standard conditions the tensile strength (F) to be prescribed for the geosynthetic to be used has to be calculated as follows:

$$F = F_b \times r_k \times r_m \times r_h \times r_n = F_b \times 2 \times 1.1 \times 1.05 \times 1.33 = 3.1 \times F_b$$

In case no seam is present, $F = 2.3 \times F_b$.

F_b is the calculated strength derived from the soil-mechanical calculation, thus including the soil-mechanical safety factor.

4.3 *Types of reinforcement and anchoring*

Types of reinforcement
In the literature two types of reinforcement can be distinguished: Stretch reinforcement and membrane reinforcement [3].

A stretch reinforcement is laid in the same direction as the acting force. An example is a geosynthetic reinforcement below an embankment on a natural ground with low bearing capacity (Fig. 3.1). During sliding a movement occurs between two masses of ground at an angle with the geosynthetic which has been anchored in both ground masses. By this movement the geosynthetic is forced in the direction of the sliding at the place of the crossing of the slip circle with the geosynthetic; Section 4.5 gives an elaborated example.

With a membrane reinforcement a force vertical to the plane of the reinforcement is applied. The geosynthetic absorbs the force by forming a domed shape (Fig. 3.2); Section 4.6 gives an elaborated example.

A steep wall construction with a grid.

Anchoring

The stress occurring in the reinforcement has to be transferred to the surrounding ground masses to make it possible for the geosynthetic to act as a reinforcement. This kind of load transfer is called anchoring. Anchoring in the soil takes place through friction between the geosynthetic and the surrounding soil (as with a grout anchor). In case of a flat anchor the requested anchoring length depends on:

– The effective soil-grain pressure (σ') at right angels to the geosynthetic, to be derived from the soil mechanical calculation;

– The angle of friction (δ) between the soil and the geosynthetic; tan δ for rough woven fabrics and grids may be Rtan φ where φ is the angle of internal friction of the surrounding soil and R a factor depending on the kind of geosynthetic; R varies between 0.7 and 0.9.

The anchoring length is calculated as follows:

$$L_a = \frac{F_a \times f_s}{n \times \sigma' \times \tan \delta}, \text{ in which:}$$

L_a = Anchoring length in meters
F_a = Anchor force in kN/m
σ' = Soil grain pressure in kN/m^2
tan δ = Tangent of the angle of friction between soil and geosynthetic
n = Factor depending on the contribution in friction of one or two sides of the geosynthetic. Hence the factor is 1 or 2.
f_s = Safety factor to be chosen between 1.5 and 2.

0 = No contribution to the friction resistance from this side
+ = Contribution from this side

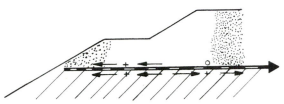

Figure 4.5. (a and b) Anchoring.

The anchoring length L_a may be divided over both sides of the geosynthetic. If the soil placed on top only acts as a dead load and can freely glide together with the geosynthetic, the friction of only one side has to be taken into account (Figs 4.5a and 4.5b). The maximum friction which can be mobilized is the friction between the soil particles. There are many alternatives for the above-presented principle of anchoring. The most frequently applied alternatives are to wrap round or to dig in the geosynthetic. It is difficult to calculate the needed anchoring length for the wrapped situation. For every single situation the loading has to be analyzed. The anchoring length can be calculated by the above-presented formula step by step. It has to be noted that the calculated anchoring length in most cases is shorter than the length needed for application reasons.

4.4 *Design considerations*

In the design period of a soil reinforcement all the forces have to be analyzed for all building phases and all loading situations, in which all time-depending factors, particularly the strength, have to be taken into account as well. The appearing forces are elaborated from soil-mechanical calculations at which several collapsing mechanisms have to be studied. From these studies it may be concluded if a reinforcement is required.

In this connection the following questions are to be answered:
– Do the leading forces occur during the whole lifetime of the construction or only during the building phase?
– Is the subsoil getting more bearing capacity in due time and will the geosynthetic become superfluous?
– May creep lead to reduction of the strain in a geosynthetic and can ongoing creep lead to the collapse of the construction?

4.5 *Design calculations for an embankment on a natural ground with low bearing capacity*

For embankments on subgrades with low bearing capacity four failure modes can be distinguished (Fig. 4.6).
  – Failure of the bearing capacity of the subsoil.
  – Internal failure of the embankment structure (embankment edge failure).
  – Failure of the embankment and the subsoil by slip circle rotation.
  – Failure of the subsoil by squeezing out.
For the following description of the failure phenomena it is assumed that the geosynthetic is applied as a stretch reinforcement.

Bearing capacity failure (Fig. 4.6a)
In the situation of an embankment on a thick layer of badly permeable soil the increase of loading is initially borne by the water pressure. The grain pressure and hence the resistance against sliding does not increase instantly and may be insufficient to bear the embankment. The application of a geosynthetic reinforcement does not affect the bearing capacity of the subsoil in this case.

Embankment edge failure (Fig. 4.6b)
The slope angle of an embankment has to be chosen in such a way that the internal stability is assured. In case an embankment is built in a short time on a badly drained and unconsolidated subsoil but with enough bearing capacity, the embankment tends to slide in the direction of the arrow in Figure 4.6b. In this situation the geosynthetic may offer a counteracting force. This force is equal to the active soil pressure. The active soil pressure is transferred to the geosynthetic by friction between geosynthetic and the material of the embankment.

Failure of the embankment by slip circle rotation (Fig 4.6c)
From observation one has learned that sliding areas can appear with a circular shape in case of embankments, built with homogeneous material on a homogeneous subsoil. For calculation purposes these sliding areas may be estimated by a calculation according to circular slip circles. For this situation Bishop has formulated a calculation method [3].

When a geosynthetic is applied the resistance giving M against the sliding soil is enlarged by the additional resistance offered by the geosynthetic at the crossing of the geosynthetic with the sliding circle.

This additional resistance giving moment (ΔM) is determined by the calculation of the force which occurs in the crossing of the geosynthetic with the sliding area in the direction of the geosynthetic. This can be calculated as follows:

$\Delta M = F_{max} \times a$, in which:

F_{max} = Admissible permanent tensile force by unit of width in the geosynthetic.

a = Lever arm of the force F to the slip circle centre (see also Fig. 9.2).

The safety factor according to the calculation method of Bishop can be derived at
as follows:

$$f_s = \frac{M_{restoring} + \Delta M_{restoring}}{M_{disturbing}}$$

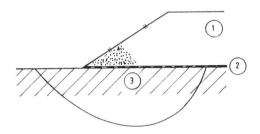

1 Embankment
2 Geosynthetic
3 Subsoil

a. Exceeding the bearing capacity of the
subsoil.

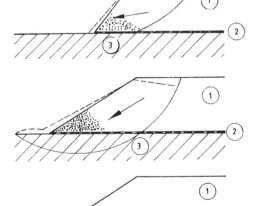

b. Embankment edge failure.

c. Failure with a slip circle rotation.

d. Squeezing.

Figure 4.6. Failure modes of an embankment on a natural ground with low bearing capacity.

As the most unfavourable sliding circle has to be calculated, calculation by computer is the only practical method.

With reference to [3] the calculation method of Bishop is in accordance with the results of practical tests. In Chapter 9 a calculation example is elaborated according to the method of Bishop. It has to be noted that the effect of reinforcement can only appear if the strain in the geosynthetic can be transmitted to the surrounding soil in case of anchoring. See for this item also Section 4.3.

Failure by squeezing (Fig. 4.6d)
Squeezing is the lateral migration of a soft layer between two relatively stiff layers (effect of an ice-cream wafer). To determine the possibilities of squeezing after application of a geosynthetic the calculation method mentioned in [3] is advised. The influence of a geosynthetic on squeezing is not yet sufficiently determinated. It is known that during horizontal deformation of the subsoil the geosynthetic gets under strain.

In the calculation example of Chapter 9 this phenomenon is discussed.

Road foundation on a subgrade with low bearing capacity: Woven fabric, wood flakes, nonwoven, light weight aggregate.

4.6 *Design calculation for a road base on a natural ground with low bearing capacity*

Analysis of the working of a geosynthetic between the road base of a road and the natural ground indicates that apart from the function of membrane reinforcement also stretch reinforcement contributes to the total bearing capacity of the road.

The tensile force in a geosynthetic is mobilized by rutting and deformation of the subsoil.

During deformation of the subsoil the geosynthetic is stretched in all directions as a result of the loading by traffic. Bij using vertical components of the tensile strain in the geosynthetic the loading can be spread over a larger area (Fig. 4.7).

The deformation of the geosynthetic and the subsoil goes on until the subsoil bears the loading without loss of stability. The geosynthetic reinforcement contributes in a constructive sense, but it is not able to prevent larger deformation in the subsoil. Yet it has been learned from comparative investigations that, if a geosynthetic reinforcement is applied, the thickness of the sub-base can be diminished [4.5]. Besides, the geosynthetic acts as a separation layer between two layers of granular material and so it saves a considerable amount of granular material, which otherwise is pressed in the weak subsoil.

Design methods for road constructions with geosynthetics are still being developed. Qualification and quantification of subsoil properties and of sub-base materials give considerable problems. Only recently a design method has been presented in which the dispersion of loadings is connected with the E-modulus of the geosynthetic and the rutting.

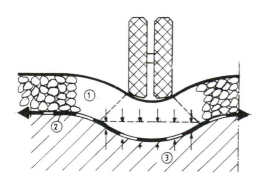

1 Road base
2 Geosynthetic
3 Natural ground.

Figure 4.7. Reinforcement of a road base ground

4.7 Design method for a steep slope bank

Geosynthetics in steep slope banks have the function to prevent the sliding of the bankedge.

In most cases the geosynthetic is wrapped around at the edge and acts as a soil container and only stems the wrapped soil layer. It has been proved in practice that the length of the wrapped end has to be at least three times the thickness of the soil layer.

Three collapsing mechanisms can be distinguished [2]: Failure of the bearing capacity of the subsoil, internal failure and failure by slip circle rotations.

Failure of the bearing capacity of the subsoil (Fig. 4.8a)

The bearing capacity of the subsoil beneath a steep slope can be calculated according to the calculation method for bed plate foundations. The bearing capacity not only depends on soil characteristics but also on the form and dimensions of the foundation surface, the slope angle and the eccentricity of the loading. Brinch Hansen has developed a calculation method for a bed plate foundation [6].

Also the calculation method based on slip circles, set up by Bishop, may be used. In most cases the bearing capacity of the subsoil is sufficient, so that this collapsing mechanism is not a critical one.

Steep slope construction with temporary retaining wall during realization.

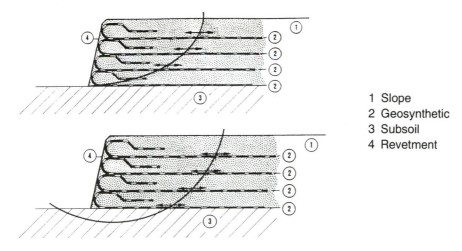

1 Slope
2 Geosynthetic
3 Subsoil
4 Revetment

Figure 4.8. (a and b). Reinforcement of a steep slope: a. Failure of the bearing capacity of the subsoil. b. Internal failure of the embankment as a whole and per layer.

Internal failure of the embankment (Fig. 4.8b)

The internal stability of a reinforced slope bank is based upon the equilibrium between two consecutive layers and hence dependent on the layer thickness, the anchoring length, the grain pressure and the properties of the soil in the embankment.

Due to reasons of execution, particularly the temporary retaining constructions and the possibilities of compaction, the layer thickness is restricted to about 1 m.

Computer computations are used to determine the critical slip circle and the requested anchoring length of each separate layer in the steep slope construction.

The strain which occurs in a reinforcing layer depends on the chosen layer thickness and increases with increasing thickness. With layers of equal thickness it is possible to design the most economical construction by applying geosynthetics with various strengths. The finding of the most economical ratio between layer thickness and strength of the geosynthetic is a question of trying out.

Jewell has drafted design charts using many computer computations, which are being used in practice.

Failure by slip circle rotation (Fig. 4.8a)

Slip circle rotation is a normative factor for the stability of the slope of the embankment dependent on the slope angle in case of high ground water level in the embankment and when the properties of the soil in the embankment are less favourable. The design graphs of Jewell can be used to calculate the safety factor against sliding [7].

4.8 *Remaining aspects*

Geosynthetic assembly

Jointing systems which are without strain can be made with a loose overlap of circa 1 m or with a simple staple or overlap seam.

Jointing systems under stress have to be avoided as much as possible. They are always weaker than the original, not-connected geosynthetic. In particular cases in which heavy forces occur in the main direction as well as at a right angle to the main direction it is usual to apply two layers, one in each direction.

If it is not possible to avoid a joint in an assembly under strain, a loose overlap is out of question, unless the overlap is as large as the total required anchoring length.

Woven fabrics can be sewed. The most usual seams are mentioned in Table 4.3 with their limitations.

Staple seams can be made in situ using a special adapted sewing machine. Overlap seams cannot be realized in situ because for this purpose sewing machines with a long free arm and several needles are requested. These sewing machines are very vulnerable in situ. At the terminations of a geosynthetic no special provisions are necessary along a selvage.

A not-selvage is provided with a hem by a sewing machine to prevent ravelling.

Damage

Under normal conditions and careful application no damages are be expected.

Attention has to be given to possible damages by lorry traffic and especially by tracked vehicles. Sometimes, depending on the location, there is a risk of vandalism.

Designing the geosynthetic the effect of possible damage has to be taken into account (see Section 4.2).

Table 4.3. Most frequently applied seams.

Description	Staple seams		Overlap seams	
	Single	Wrapped	Single	Wrapped
Shape				
Strength of the seam in % of the strength of the woven fabric	25-50	30-60	60-80	60-80
Soil tightness in finegraded soils	Doubtful	Guaranteed	Doubtful	Guaranteed

Secondary properties

Geosynthetics with a main function of reinforcement have mostly to be water permeable. Sometimes soil tightness is requested when several types of soil have to be separated. If based upon the main function a chosen geosynthetic does not meet the requirements of a secondary function; a combination can be made with another geosynthetic. Of course the influence of the second geosynthetic on the main function of the first one has to be taken into account.

Closing remark

Operation restrictions and execution conditions are discussed in Chapters 7 and 8.

4.9 *Summary*

Strength and stiffness are the two distinctive properties of soil reinforcement with geosynthetics. Changes in temperature, alteration, creep and damage may have a great influence on the admissible stress. Hence only polyester woven fabrics and polyester grids with a high E-modulus are suitable as reinforcing material.

Brook-bank revetment with a reinforced bituminous membrane.

Depending on the manner of loading two types of soil reinforcement can be distinguished:

 – Stretch reinforcement with loading in the plane of the geosynthetic;
 – Membrane reinforcement with the loading vertical to the plane in which the geosynthetic is lying.

Soil reinforcement with a geosynthetic is only possible if the geosynthetic is anchored.

The transfer of forces is realised by friction between the geosynthetic and the surrounding soil.

Designing a soil reinforcement time-depending factors have to be taken into account (loading and strength).

For three total different types of soil reinforcement design considerations and calculation methods are presented. Remaining aspects of the application of geosynthetics as a soil reinforcement concern: joints and overlaps, and secondary properties related to water-permeability and soil-tightness.

Sometimes a secondary function requires the application of a second product (composite).

CHAPTER 5

Filter and separation function

5.1 *Introduction*

In Section 2.2 two examples are presented concerning filter and separation function:
- River bank protection in which a geosynthetic performs a function between soil and water;
- Road foundation/sub-base, in which the geosynthetic separates sub-base material and natural ground.

For the filter function two properties are of importance:
- Soil tightness (retaining of soil particles);
- Water-permeability.

If the main function is to separate two types of soil, it is called 'separate function'.

In Section 5.2 the function properties are discussed which a geosynthetic must have to function as filter or separation. The properties of pore dimensions, permeability and the methods to determine these functions are presented.

In Section 5.3 the design is presented. Design criteria are discussed in which special attention is given to grain size distribution and grain shape in relation to the pore dimensions and the permeability of the geosynthetic.

In Section 5.4 design rules are presented.

In Section 5.5 remaining aspects are mentioned.

In Section 5.6 a summary is given.

5.2 *Functional properties*

Soil tightness and water-permeability are in principle conflicting material requirements which have to be fulfilled simultaneously. The soil tightness requires a reduction of the pore dimensions, on the other hand the permeability requires large pore apertures.

Application of a nonwoven as river-bank protection with a filter construction covered with rip-rap.

Soil tightness

The soil-tightness of a geosynthetic is marked by the so-called soil-tightness number $O(p)$ in μm. $O(p)$ is determined by the ratio of a characteristic pore dimension of the geosynthetic and a certain characteristic grain diameter d_m of the bank material to be protected.

$O(p)$ has been defined as the main grain diameter of the sand-fraction, of which $p\%$ (mass/mass) remains on and in the filter after five minutes of dry sieving conform the method described in the Netherlands design standard NEN 5168.

To determine the soil-tightness number the geosynthetic is used as a sieve for a sand-fraction with the fraction limits d_{min} and d_{max}, corresponding with two successive standardized sieves; d_m is the average of d_{min} and d_{max}.

After 5 minutes of sieving of the sand-fraction through the geosynthetic a sieve-rest $p(\%)$ remains in and on the geosynthetic. These sieve rest percentages of 90 and 98 are used in design rules as characteristic pore dimensions $O(90)$ and $O(98)$.

Because the result of one sieving rarely gives the requested sieve percentage more sievings have to be performed and by interpolating between the results of sieving of two close fractions the requested figures can be derived at. Example:

Fraction d_{min} (μm)	d_{max} (μm)	d_m (μm)	Sieve-rest p (%)
250	300	275	85
300	355	328	93

With $p = 90$ there belongs a $d_m = 328 - 3/8 \times (328\text{-}275) = 308$ μm.

In conformity with NEN 5168 the average of the results of sieving of 5 samples has to be determined. In the product catalogue of the Netherlands Geotextile Organization $O(90)$ is mentioned as the characteristic pore dimension.

When the soil-tightness of a geosynthetic is compared with the actually present soil particles the characteristic value of the grain size of the soil d_p in μm is determined, in which d_p is the sieve number which is passed by $p\%$ of the soil sample. Hence: At $d_{90} = 50$ μm, 90% of the soil sample passes a sieve with a mesh of 50 μm. Normally d_{90}, d_{85}, d_{60}, d_{50}, d_{15} and d_{10} are used.

Water-permeability
When describing the water-permeability of a construction with geosynthetics the expressions permittivity and transmissivity are used. Permittivity is the volumetric flow rate of water perpendicular *on* the plane of the geosynthetic. Transmissivity is the volumetric flow rate of water *in* the plane of the geosynthetic, especially of importance in drainage.

The water-permeability in terms of permittivity (ψ) can be defined as follows:

$$\psi = \frac{v}{\Delta h} \, [s^{-1}]$$

From this formula the following can be derived:

$$\psi = \frac{k}{e} \, [s^{-1}]$$

where:

ψ = Permittivity in s^{-1},
v = Filter velocity m/s,
Δh = Hydraulic head difference across the geosynthetic,
k = Coefficient of permeability for laminar flow (m/s),
i = The hydraulic gradient across the geosynthetic,
$i = \Delta h/e$,
e = The thickness of the geosynthetic (mm).

The permittivity keeps a constant value as long as the flow through the geosynthetic remains laminar. The temperature and the filter velocity affect the permittivity.

If in a product information bulletin the permittivity is presented, this is valid for a standard temperature of $T_s = 10°C$ and a standard filter velocity of $v_s = 10$ mm/s. In this situation the hydraulic head difference is called hydraulic head difference Δh_s or loss of pressure height P_s. This value P_s is also often mentioned in product information bulletins and it is easy to derive at from the above mentioned formula of the permittivity:

$$P_s = \Delta h_s = \frac{10}{\psi} \, [mm]$$

According to the law of Darcy the water-permeability of soil is expressed by the k-value: $v = k \times i$. For geosynthetics the hydraulic gradient is:

$$i = \frac{\Delta h}{e}$$

therefore

$$k = \frac{v}{i} = \frac{v \times e}{\Delta h} = \psi \times e$$

or the permittivity is:

$$\psi = \frac{k}{e}$$

The Netherlands design standard NEN 5167 presents how the permittivity has to be determined.

Remark: The product information bulletins do not always follow the above given systematics.

Honey cell structure of a nonwoven to prevent slope erosion.

5.3 *Design considerations*

Soil-tightness

Transport of soil particles within a grain structure is possible when there is a driving force (groundwater pressure, fall within the ground).

The intention in most cases is to prevent the transport of small-sized soil particles in the subsoil and therefore the term soil-tightness is used and not the term space for transport or pore-volume (in case of transport of water the term pore-volume and water-permeability is used).

The relation between pore-magnitude and grain-diameter can be characterized by: Pore-diameter ≈ 20% of the grain-diameter.

Just as for the characterization of the ability of a grain structure with regard to the transport of soil particles, for geosynthetics, too, the term soil tightness is used.

In a theoretical case the soil is composed of spheres of one size diameter. All spheres can be retained, if all apertures in the geosynthetic are smaller than the diameter of the spheres.

Usual the soil consists of particles with different diameters and shape, which is reflected in the particle-size distribution curves. Smaller particles can disappear straight across the geosynthetic by ground water current. In this case the retained soil structure can function as a filter; see Figure 5.1.

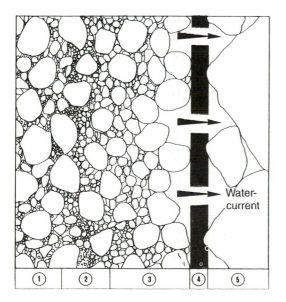

1 Original soil structure
2 Filter zone
3 Filter cake
4 Geosynthetic
5 Revetment

Water-current

Figure 5.1. Schematic presentation of a filter with a soil retaining layer.

The better the soil particles are distributed (the smoother the curve), the better the soil tightness of the soil structure is effected. Smaller soil particles get stuck into the spaces between larger ones and the soil structure prevents the flow of fine particles. When certain particle-size fractions are lacking, the soil structure is not stacked very well and cavities develop through which erosion can occur. The displacement of soil particles not only depends on the soil tightness but also on the hydraulic gradient in the soil structure.

In order to judge the risk of wash-out of soil particles through the geosynthetics some aspects have to be considered.

a) Internal stability of the soil structure

In case of a loose particle stacking of the soil many small soil particles may pass through the geosynthetic before a stable soil structure is developed near the geosynthetic.

A soil structure is stable if $d_{60}/d_{10} < 10$.

In case of vibration, for instance caused by waves or by traffic, the stable soil structures can be disturbed. To meet this situation the subsoil has to be compacted in advance and a good junction between geosynthetic and subsoil has to be guaranteed.

Application of a nonwoven as separation layer in a drainage system.

b) Stability of the soil structure at the transit of grains to the geosynthetic
As a result of fluctuating loadings the filter-working can be disturbed near the junction of the subsoil with the geosynthetic. This phenomenon can appear for instance with turbulent flows, where strong pressure fluctuations occur caused by ship and wind waves. This kind of disturbance of the soil structure cannot be prevented and therefore a geosynthetic has to be chosen with a small pore-size. However, the water-permeability may not be diminished unacceptably.

c) Channel forming between the geosynthetic and the edge of the soil structure
Above all the wash-out of soil particles has to be prevented, because as a result of it settlement and loss of stability can be caused. For the same reason the development of channels and holes below the geosynthetic must be prevented. Therefore, the geosynthetic has to follow the roughness of the subsoil.

d) Stability of soil structures on both sides of the geosynthetic
In case a geosynthetic is applied as a separation layer in a road subbase, this layer is situated between the natural ground and the sub-base.
Through traffic load the soil structure is deformated and fine soil-particles can penetrate in the sub-base. The appearing forces are usually extremely high, although of short duration; yet enough to transport small loose soil particles. These small soil particles may form impermeable slices in the sub-base. Hence water may accumulate above these slices causing frost heave and sub-base instability owing to thaw. Application of a geosynthetic with a pore size adapted to the smallest grain size in the subsoil prevents these harmful phenomena.

Waterpermeability
To prevent the forming of water-pressure in the construction, causing loss of stability, the geosynthetic has to be water-permeable. One has to strive for the increase of water-permeability of a construction in the direction of the water-current. In case of a river-bank protection it means that the permeability of the geosynthetic has to be larger than the permeability of the soil on which the geosynthetic has to be applied. In case of a road foundation the geosynthetic is usually applied on an impermeable layer of clay or peat. The water-permeability of woven fabrics and nonwovens in all these situations is much larger than that of the natural ground. The water-permeability of woven fabrics and nonwovens may decrease in course of time owing to the fact that fine soil-particles, which are transported by the groundwater-flow from the subsoil, migrate into the pores of the geosynthetic (clogging).

This clogging phenomenon is only rarely observed because of the forming of a natural filter in the subsoil [8.9]. In case of river-bank revetments clogging may occur, caused by mud particles from the river-water. Practical tests have proved that building of pressure below the geosynthetic is negligible. The risk of blocking is present if the following three circumstances appear at the same time (what seldom happens):

 – The aperture in the geosynthetic is very uniform;
 – The grain-size of the subsoil is very uniform;
 – $O(90)/d_{90}$ is between 0.5 and 1.0.

In this case every aperture in the geosynthetic can be closed, and as a result pressure building appears below the geosynthetic and the grains cannot be removed from the apertures in the geosynthetic.

Clogging may also appear very slowly. Here three factors play a role, which, separately, may cause clogging:
 – Biological growth;
 – Deposition of fine mud-particles from the river-water;
 – Chemical deposition in the geosynthetic by water containing a large amount of iron or chalk.

To prevent this kind of clogging the pore-size of the geosynthetic has to be chosen as large as possible; but, of course, this pore-size has still to meet the requirements for soil-tightness.

5.4 *Design rules*

Soil-tightness
With respect to the soil-tightness some general requirements can be formulated. However, in many situations additional requirements will be necessary dependent on the local situation. Based upon practical experience the basic requirements are presented in Table 5.1.

The internal stability of a grain-structure is expressed in the ratio between d_{60} and d_{10}. As a rule this value has to be smaller than 10 to guarantee sufficient stability.

Table 5.1. Design requirements for geosynthetics with a filter and separation function.

Description	Filter function
Stationary loading	$O(90) < 2 \times d_{90}$
Cyclic loading – Stable soil-structure	$O(98) < 2 \times d_{85}$
Cyclic loading – Non-stable soil-structure – No effects by wash-out	$O(98) < 1,5 \times d_{15}$
Cyclic loading – Non-stable soil-structure – Inadmissible effects by wash-out	$O(98) < d_{15}$

Water-permeability

As a design criterion one can hold that the water-permeability of a geosynthetic has to be greater than that of the soil at the side from which the water-flow comes. As a rule one can keep to:

$$k_{\text{filter}} = k_{\text{soil}} \times \text{factor}$$

If a geosynthetic is permeable with a factor of 10 more than the subsoil, overpressure will not arise, neither below the geosynthetic, nor in case of reduced permeability caused by clogging or blocking. If there is no danger for clogging or blocking a factor of 2 or 3 is sufficient.

5.5 Remaining aspects

Seams and overlap

Wrapped, sewed seams are reliably soil-tight, also for fine graded soil-types. Single sewed seams may be unreliable when applied on fine graded soil-types; see also Table 4.3.

If sewed seams are not possible in practice, an overlap has to be applied. A guideline for an overlap dimension is:
- Execution under dry conditions: at least an overlap length of 0.5 m;
- Execution in wet conditions: at least an overlap of 1.0 m.

Remarks

a) In case of a combination of reinforcement and filter-function, seams and overlaps are undesirable.

b) When unrolling the geosynthetic, araising of wind has to be prevented and no soil may appear between the two layers in the overlap. Timely ballasting is consequently necessary.

c) When a filter-construction of large dimensions like blockmats, mattresses and willow mattings is applied, a particular, ballasted overlap construction is used. Figure 5.2 presents an example of such an overlap construction.

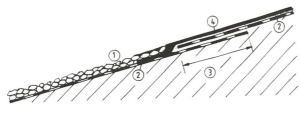

1 Graded quarry stone
2 Geosynthetic
3 Overlap
4 Open stone-asphalt concrete

Figure 5.2. Particular overlap construction.

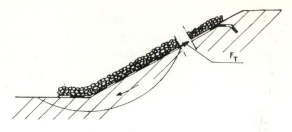

Figure 5.3. Sliding of a slope.

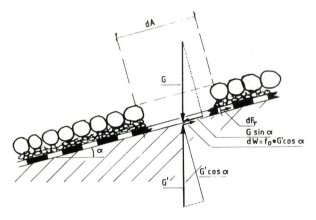

Figure 5.4. Sliding of the geosynthetic and the revetment across the subsoil.

Secondary function

Particularly when filter-constructions are applied on slopes, the tension-force in the geosynthetic has to be examined. In two cases tension-forces will arise:

   a) A subsoil failure; see Figure 5.3.

   If the geosynthetic is anchored high in the slope and a slope failure appears, a tensile force F_T is developed in the geosynthetic. This force can be enormous. The geosynthetic of a filter-construction is not the material to prevent this kind of instability.

   b) Geosynthetic and revetment slide by insufficient friction between geosynthetic and subsoil; see Figure 5.4.

   The loading G on a slope-part dA has a force-component down along the slope of $G \times dA \times \sin \alpha$.

   The friction will prevent the sliding: $dW < = f_o \times G' \times dA \times \cos \alpha$.

   f_o is here the coefficient of friction between the underside of the geosynthetic and the subsoil and G' is the reaction on the loading G.

If $dW < G \times dA \times \sin \alpha$, a tensile force dF develops at the upper side of slope-part dA. The contribution of dA to the total tensile force in the geosynthetic, under the condition that the geosynthetic has been anchored at a higher part of the slope, is consequently: $dF_T = G \times dA \times \sin \alpha - f_v \times G' \times dA \times \cos \alpha$. When the filter-construction has been calculated correctly, the water-pressure against the underside of the filter will be small and in this case the grain-pressure of G', which has to deliver the friction, will be equal to the pore-pressure of G. After some conversion it appears that at normal values of f_o and α no tensile force F_T will develop in most cases. ($f = \tan \delta = 0.6$-$0.9 \tan \varphi$, where δ = the angle of friction between geosynthetic and subsoil, and φ = the angle of internal friction of the subsoil).

Only in case of water-overpressure below the geosynthetic, for instance as a result of a rapid change of the river water table, the friction may become too small because of a decreased grain-pressure. In this situation a drainage below the geosynthetic and a good water-permeable geosynthetic and revetment prevent over-pressure.

Particular damages
Filter constructions with geosynthetics for riverbank and river-bottom protection are vulnerable during application.

The risk of damage caused to the geosynthetic by falling stones should not be underestimated. Damage to the geosynthetic can be prevented by:
   – The application of a load-spreading layer of gravel or light stones (maximum 10 to 60 kg);
   – Reducing the height of the fall of rockfill or heavier stones, by placing the dumping vessel or crane bucket as near to the mattress as possible.

A special situation of damage occurs when by insufficient friction the revetment slides across the geosynthetic; see Figure 5.5.

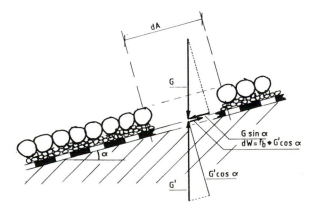

Figure 5.5. Sliding of the revetment across the geosynthetic.

Separation of sub-base material and natural ground for a road construction.

With f_b as friction coefficient between upper-side of the geosynthetic and revetment the friction is: $dW = f_b \times G' \times dA \times \cos \alpha = f_b \times G \times dA \times \cos \alpha$; for G' is G, regardless of the slope-protection construction being situated above or below the water-level and regardless of the possibly present water-overpressure below the geosynthetic.

Derivation learns that sliding does not appear when $f_b > \tan \alpha$. In most cases sliding appears during execution by negligent stone-dumping, but seldom afterwards. When it happens, the geosynthetic will be damaged by the sliding revetment.

Final remark
Application restrictions and execution terms are discussed in Chapters 7 and 8.

5.6 *Summary*

Soil-tightness (retaining of soil-particles) and water-permeability (permittivity) are the two functional properties of the geosynthetic as filter or separation layer.

The pore-size $O(p)$ of a geosynthetic, the grain-size diameter d_p of the subsoil, the variation in both and the ratio $O(p)/d_p$ are the determining properties for a good working geosynthetic as a filter. For the soil-tightness easy design rules are formulated for the ratio $O(p)/d_p$.

The water-permeability of a geosynthetic is unambiguously described by defining the permittivity (ψ) as the ratio between the standard current velocity in the geosynthetic (v_s) and the hydraulic head difference across the geosynthetic Δh_s.

Geosynthetics, applied as a filter, have to be much more water-permeable than the soil at the side from which the groundwater-flow streams into the geosynthetic.

In case the subsoil is compacted insufficiently or when cyclic loadings appear, there is a great chance for wash-out through the filter and below the filter. Therefore, during execution, special attention must be given to joints and overlaps. The water-permeability of a geosynthetic may decrease by clogging and blocking. If there is any chance of this, the most suitable geosynthetic has to be carefully selected, if necessary based upon soil analyses.

Remaining aspects of the application of geosynthetics concern problems near seams and overlaps, with slope revetments under tensile tension and the chance of damage during execution (transport, sinking and stone dumping).

Vertical drainage of an embankment with a geosynthetic.

CHAPTER 6

Screen function

6.1 Introduction

In Section 2.2. three examples are given of the screen function of geosynthetics:
- Sealing of a reservoir;
- Road in a cutting below groundwater table;
- Isolation of a landfill.

The main function of a screen is to be impermeable for liquids. Only membranes and other geosynthetics with membrane properties are suitable, as can be learned from Table 3.2. As a basic material mostly polyethylene is used and, if permitted, polyvinylchloride. Besides membranes composites like reinforced asphalt mats, rubber membranes or laminates of synthetic materials are also used.

In Section 6.2 the function properties are discussed which geosynthetics must have in order to fulfil the screen function (water-impermeability).

In Section 6.3 the design is discussed.

Section 6.4 deals with the remaining aspects. Here the accent is on seams and joints and how these can be made in situ. Also the secondary function and particular risks of damaging are discussed.

In Section 6.5 a summary is given.

6.2 Functional properties

Characteristic of geosynthetics in a screen function is the resistance against liquid percolation called liquid tightness. In product information bulletins the term 'permeability' (not 'impermeability') is often used. The permeability is determined in a laboratory.

A good new geomembrane has a permeability of 10^{-13} to 10^{-15} m/s (NEN 5167). In practice the thickness of the membranes differs from 0.5-2.5 mm for membranes and from 1.5-6 mm for reinforced asphalt mats. In comparison, a good compacted sealing layer of clay has a permeability of 10^{-10} m/s.

Dependent on the consequences of the exceeding of criteria, requirements can be called for regarding the liquid tightness. In this regard, for instance, in order to prevent salt percolation less severe requirements are demanded from a seepage screen than from a screen to prevent the seepage of petrol from a leaking tank of a filling station. From this example it is also clear that, depending on the local situation, requirements have to demanded from the chemical resistance (see also Section 7.3).

6.3 *Design consideration*

A membrane is applied on a plane surface of sand or clay. Some roughness in this subsurface layer can be absorbed by the stretch of some percentages of the geomembrane. If settlement occurs, the deformation of the subsoil may be considerable and the risk exists that the membrane tears or is stretched in such a way that the liquid tightness gets lost.

Geotechnical investigations will give enough information regarding the risk of settlement in order to take measures.

Road in a cutting below groundwater table; application of a membrane, tunnel De Noord in the Netherlands.

Pressure differences between both sides of a screen may be caused by permanent or by varying gas or water-pressures. Water-pressure differences can be distinguished in slow fluctuations, like seasonal fluctuations in the water-table, and fast fluctuations, as occur by wash of the waves. Besides, prolongated pressure differences may occur by gas-forming in the subsoil.

The whole construction has to be resistant to these pressure differences. The resistance is secured by ballasting the screen with soil or graded quarry stone. To reduce or to prevent pressure building at one side of the screen, a drainage can be applied at that side. This drainage layer can detect and transport the seepage-water in case of demolition.

6.4 *Remaining aspects*

Seams and joints

Slips of membranes are connected by glueing or melting of overlaps. Membrane joints are also weak points and thus joints made in situ have to be limited as much as possible by making them in a factory under controlled circumstances. In practice the transportation possibilities are the limiting factors.

In a factory joints can be made by high-frequency welding, ultrasonic welding or glueing. These technologies have to be performed with special equipment and very carefully, sometimes even in conditioned rooms. These kind of joints result in very good connections but they cannot be performed in situ. The joint-width is 20-30 mm for single joints and two times 10-20 mm for double joint (two single joints next to each other).

Construction of a landfill.

The most frequently occurring welding technologies for membranes in situ are:

– Hot-air melding; here the overlaps are heated with hot-air and then pressed together;

– 'Heizkeil' welding; here a heated wedge is drawn between the overlap of the slip to be welded, and then pressed together;

– Extrusion welding, a system of hot-air welding, in which at the same time glue or another material is added.

Because the liquid tightness is the functional property of membranes, the joints have to be liquid-tight as well. This is also valid for joints made in situ. These in situ-made joints have to be controlled and tested directly after welding and along the total welding length.

For this purpose two methods are available:

– At a double joint air is blown in the space between the two welds; leaks can be detected in the same way as with bicycle tubes;

– At a single joint a brass-wire is added to the joint. After finishing the welding operation the brass-wire is brought under voltage; then a detector is moved over the weld and where there is a leakage, sparks will leap over from the brass-wire to the detector.

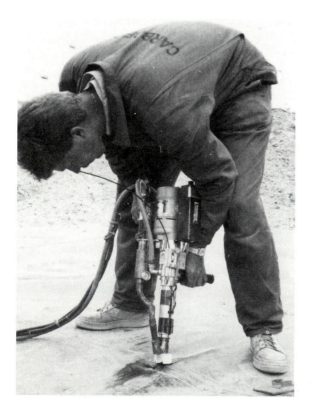

Welding of membranes.

A clear site for welding activities on a landfill.

Asphalt mats (not applied in landfills) are placed with an overlap of half a meter and pressed together after hot asphalt bitumen has been poured over the underlying mat. For a good result it is essential that the crew has experience in the technology to be applied and that the technology has been tested previously at a laboratory.

Besides the applicability of the technology also the work and weather conditions are of importance. At a muddy site, which cannot be kept sufficiently dry and where building rubbish lies around, no joints of sufficient quality can be expected. A clear site is essential for a good performance.

Secondary function

With a screen construction on a slope tensile force may occur, which has to be examined; see Section 5.5 for a complete explanation for the symbols used.

A tensile force can be thought of in two cases:

a) The resistance against sliding in the subsoil is limited, for instance as a result of the enclosure of small loam or clay layers, deposited during the construction of the pit. In this case a slip circle will develop in the slope; see Figure 6.1.

b) As a result of the water-permeability and sometimes also as a result of the gas-tightness of the membrane a water and gas-overpressure σ_{wo} may develop in the slope below the membrane, resulting in a decrease of the friction resistance against sliding of the membrane plus ballast over the subsoil; see Figure 6.2.

In both cases a tensile force F_T develops in the membrane. In case the membrane is firmly anchored at the upper-side, this secondary loading leads to the exceeding of the break-elongation limit of the membrane, as a result of the membrane tears.

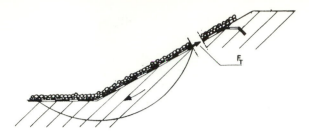

Figure 6.1. Sliding of a slope.

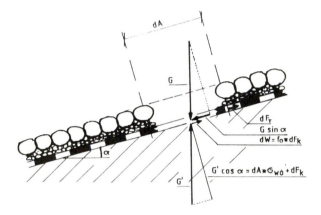

Figure 6.2. Sliding of the geosynthetic plus ballast over the subsoil.

In the above-described situation the functions have to be separated by combining the membrane with a woven fabric of sufficient strength or by choosing a woven fabric coated with a watertight layer for instance of PVC.

Sliding of (parts of) the slope is determined by soil-mechanical factors.

Sliding of the membrane plus ballast across the subsoil can be prevented by applying a drainage layer below the membrane so that a water or gas-overpressure cannot develop; see also Section 5.5.

If the risk of sliding and the anchoring have to be calculated, the value of the friction angle δ between membrane and subsoil has to be taken into consideration. For a smooth and hard material tan δ = 0.6 tan φ; for a rough surface of a membrane this value may increase to 0.9 tan φ. By printing a rough texture onto the membrane during fabrication (calendering) the greatest possible value can be obtained: $\delta = \varphi$.

Exceptional damages

All constructions with a membrane in a screen function have to be executed very carefully, particularly the kind of construction that has to prevent the pollution of the subsoil. Each damage of the membrane in landfills may lead to loss of function of the membrane.

In landfills high temperatures can arise as a result of the heating of mixed waste, through which a rapid ageing of the geosynthetic may occur. If this situation occurs, an extra sand layer is necessary to protect the membrane.

Membranes have a great coefficient of expansion, varying from 5 to $20 \times 10^{-5}/°C$. If, during application of the membrane or directly afterwards, considerable lower temperatures appear, high tension forces may occur through crimp, resulting in the tearing of the membrane, and with higher temperatures folds may occur through the extension of the membrane. This results in thin spots in the membrane, what will lead to yielding during ballasting or welding.

A special case of damage arises when ballast is sliding over the membrane through insufficient friction between both; see Section 5.5 'Particular damages'.

Concluding remark
For application restrictions and execution terms the reader is referred to Chapters 7 and 8.

6.5 *Summary*

The functional property of a geosynthetic for a screen function is the liquid tightness. In most cases a membrane of polyethylene is applied. Depending on the application and local particularities, various demands can be made.

Membrane and drainage system in a landfill.

The membrane has to be applied on a smooth and flat sub-layer. Small roughnesses in the sub-layer can be absorbed by the elongation in the membrane. Resistance can be given against permanent and cyclic loading by ballasting the membrane and by applying a drainage layer.

It is desirable to restrict the amount of seams and weldings as much as possible. If it is avoidable to make joints and wells in situ, welding technics have to be applied which have proved their reliability. Nevertheless, all weldings made in situ have to be tested on tightness. A membrane may slide through water- or gas-overpressure. The tensile strength of a membrane is often insufficient to offer resistance. In this case particular provisions are necessary. A woven fabric of sufficient strength in combination with a membrane may be a good solution. To prevent overpressure the application of a drainage layer may be the solution. To prevent damage, careful execution is requested. Great temperature differences during application have to be prevented and the risk of heating of mixed waste in a landfill has also to be avoided.

CHAPTER 7

Damage and affection

7.1 Introduction

Geosynthetics may be damaged before, during and after execution. Most damages occur during execution. Damage and affection can be distinguished in:
– Mechanical damage;
– Physical damage by the environment;
– Chemical damage by the environment.
Besides, geosynthetics may affect the environment. The following sections deal with these points.

In Section 7.2 mechanical damage is discussed.

Section 7.3 deals with the kind and consequences of physical and chemical affections of geosynthetics caused by the environment.

Section 7.4 describes the damage caused by the geosynthetic and what kind of re-use is possible.

In Section 7.5. a summary is given.

7.2 *Mechanical damage*

Mechanical damages before and during execution might be prevented by the careful handling of the geosynthetic during transport, storage and at the site. Naturally, the methods of transport and application and the execution circumstances have to be considered in the selection of the geosynthetic. For this reason the care for the geosynthetic and the application method are described in the builders specification. Some points of attention in relation to mechanical damage follow below.

Geosynthetics have to be unfolded on an almost flat surface, free of stones and other obstacles. Small, local settlement differences, lumps and potholes will always be present. To bridge these unevenesses an elongation in the geosynthetic of 5-10% is sufficient. All kinds of geosynthetics have a much greater elongation at break (Fig. 4.1).

In normal situations and with careful execution no problems will arise. At uneven spots a stress concentration will usually take place resulting in small shiftings in the subsoil leading to reduction of the stress.

In case stones lie below the geosynthetic there is a chance of punching. The punching resistance is then exceeded, which may lead to intolerable damage and loss of function of the geosynthetic. For instance a sharp stone will easily penetrate a membrane, causing leakage.

During the sinking of a mattress there is a chance of punching by falling stones. Application of a protecting layer, for instance of reed mats or a layer of gravel, and reduction of the drop-hight may be the solution.

For important constructions and before application of newly developed geosynthetics a test-case in practice is recommended to learn the collapsing mechanism and the chance of failing. Here also the execution-aspects, when there is a possibility of the above-mentioned damages, have to be studied.

2 = geosynthetic.

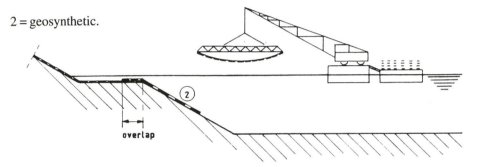

Figure 7.1. Sinking of a mattress with a special frame

Application of the block-mattress in the Eastern Scheldt storm surge barrier.

During unrolling, unfolding or sinking of a geosynthetic attention has to be paid to an equal action, so that no great and irregularly divided tension forces arise. These forces lead to irregular deformations in the geosynthetic and cause dilatation. Use of a special frame during the sinking of a mattress may prevent stress concentrations; see Figure 7.1.

Dragging of a geosynthetic over larger distances at a site may cause wear and tear. The friction may result in an increase of temperature, to such a degree that material properties may change. Addition of water may prevent increase of temperature.

7.3 *Affection by the environment*

In Table 7.1. a comparative overview is presented regarding the sensitivity of various geosynthetics to physical and chemical affection. Several kinds of affection may appear at the same time; see Section 3.5 for the most important phenomenon of 'Ageing'.

Table 7.1. Resistance of geosynthetics against affection.

Basic material	PET		PP		LDPE		HDPE		PA	
Time of exposure	Short	Long	Short	Long	Short	Long	Short	Long	Short	Long
Dilute acids	++	+	++	++	++	++	++	++	+	O
Concentrated acids	O	–	++	+	++	O	++	++	O	–
Dilute alkali	++	O	++	++	++	++	++	++	++	+
Concentrated alkali	O	–	++	++	++	++	++	++	O	–
Salt	++	++	++	++	++	++	++	++	++	++
Mineral oil	++	++	+	O	+	O	+	+	++	++
Glycol	++	O	++	++	++	++	++	++	+	O
Micro organisms	++	++	++	++	++	++	++	++	++	+
UV-light	+	O	O	–	O	–	O	–	+	O
UV-light (stabilized)	++	+	++	+	++	+	++	+	++	+
Heat, dry (up to 100°C)	++	++	++	+	O	–	+	O	++	+
Steam (up to 100°C)	O	–	O	–	–	–	+	O	++	+
Hydrolysis	++	++	++	++	++	++	++	++	++	++
Detergents	++	++	++	++	++	++	++	++	++	++

Short = during execution; Long = during usage; Degree of resistance: – = non-resistant; O = moderate; + = passable; ++ = good.

Laying a block mattress on a nonwoven.

7.4 *Affection of the environment by a geosynthetic*

Geosynthetics in civil-engineering constructions behave chemically inactive and therefore they are environmentally harmless.

In areas where drinking-water is won and in reservoirs the applied geosynthetics need a certificate. This certificate indicates that the geosynthetic concerned has been tested on aspects of health and has been approved.

In view of the increasing accent on prevention of pollution, it is recommended to inquire the potential environmental effects.

Generally speaking it can be said that pollution by geosynthetics only occurs during the execution phase, during the replacement of the geosynthetic or at the demolition of the construction.

To process the refuse of geosynthetics the following methods may be considered:

Dumping or storage
These methods are not very attractive because of space-keeping and costs of transport and dumping.

Incineration
Only the incineration of polyvinylchloride may lead to air pollution because hydrochloric-acid gas and dioxide may be formed. These gasses are particularly harmful for man and environment. Emission of hydrochloric gas can be decreased by gas-washing. Emission of dioxide may be prevented by burning PVC at very high temperatures (1400°C).

Recycling

Geosynthetics can be recovered and sometimes they are already made from recycled material. An economic condition for reuse is that enough material of one type of geosynthetic is available. Mixing of basic materials in one recycling process cannot lead to useful products, because the melting points of the various basic materials vary too much. Besides, homogeneous mixing cannot be reacted easily.

Recycled material of PE, PP, PA and PET can be regranulated and used for the production of fibres, planks, waste disposal bags and containers. The technology for reuse of synthetics is continuously being improved. More information can be obtained from the handbook 'Geotextiles and Geomembranes in Civil Engineering'.

Moreover, the following remarks are important:
– Pollution appears especially with negligent execution and demolition. Geosynthetics may be blown away, drive away and remnants may be scattered about in the surroundings. Animals may be strangled in the geosynthetics and ship-screws may jam.
– Geosynthetics do not interfere with natural processes. Only strong and thick geosynthetics prevent or retard the grow of roots.

7.5 *Summary*

Mechanical damage can be prevented by a proper choice of material and a careful execution.

Much attention must be paid to the flatness of the surface on which the geosynthetics are spread.

Danger of punching may arise when stones lie under a membrane or when stones are dumped on a membrane.

Great differences in tension and deformation lead to the forming of folds. These folds have to be prevented.

Affection of geosynthetics may arise caused through the impact of acids, alkali, oil, muck and soon. The chance of affection may determine the choice of the kind of geosynthetic.

UV-radiation and high temperatures may accelerate ageing.

There is no danger of emission of toxic materials from the geosynthetics to the environment, except some kinds of PVC.

For application in areas where drinking-water is won a certificate is requested, stating that no toxic materials will leach.

Geosynthetics which become available after site-clearing and demolition of a construction can be dumped on a landfill, burned or recycled. Special measures have to be taken to prevent emission to the environment.

CHAPTER 8

Quality assurance

8.1 *Introduction*

To determine the quality of a product to be delivered a package of requirements has to be set up in relation to a particular application. For this purpose functional requirements have to be translated to product properties. These product properties must be part of a delivery contract and must be laid down in standards and certificates.

In Section 8.2 the certification of geosynthetics is discussed.

In Section 8.3 the standards are mentioned which prescribe the test-procedure.

In Section 8.4 practical tests are described, which are sometimes requested for applications with particular risks.

In Section 8.5 terms of reference are discussed.

In Section 8.6 a summary is given.

8.2 *Certification*

To ensure a good quality review expensive and prolonged test procedures are necessary. Such tests cannot, of course, be prescribed for projects to be executed at short notice.

A quality certification offers a good possibility of being certain that the product delivered meets the requirements. The present system is directed towards a continuous assurance of quality, achieved by constant control and inspection during the production process performed by the producer and controlled by an independent organization (certifying agencies).

Important elements for quality assurance are:
- Clearly defined requirements;
- Competent personnel;
- Clearly defined responsibilities of the people involved;
- An adequate information flow (reliable, well defined, quick).

The initiative for certifying has to be taken by the producer of the geosynthetic. It is to be expected that in the near future more certificates will be asked for. The same system of quality assurance is valid for the application of geosynthetics in civil engineering works by contractors.

The professional skills of the people involved and the reliability of the execution processes will be guaranteed.

Geosynthetics (nonwoven and a membrane), used as separation in a light-weight foundation construction for an industrial application. Sequence: subsoil, membrane, polystyrene-blocks, nonwoven, light-weight aggregate.

Woven bags, filled with gravel below piers; Eastern Scheldt storm surge barrier.

8.3 *Quality assessment*

In most cases the presentation of the certificate of certified products will be sufficient.

In particular cases and for non-certified products it is necessary to verify whether the geosynthetics meet the requirements.

The buyer has to describe in the specifications or in the terms of reference how and how many samples have to be taken and where. He has to specify which test procedures have to be followed. A number of test procedures have already been laid down in standards and quidelines, others are in preparation [10].

The most important standards in relation to geosynthetics are:

Dutch standards:
NEN 5168: soil tightness
NEN 5167: resistant against flow-through of water
NEN 5132: resistant against thermic oxidation

German standards:
DIN 53857: strip tensile test
DIN 53855: measurement of thickness
DIN 53854: measurement of mass
DIN 53815: maximum stress
DIN 53834: tensile test for yarns
DIN 53668: grab tensile test
DIN 53861: burst test
DIN 54307: CBR plunger test (puncture test)
DIN 53356: leg-tear test
DIN 53859: tongue-tear test
DIN 53383: resistant against UV-radiation
DIN 53377: dimensional stability
DIN 53363: resistant to tearing
DIN 16726: perforation test

International standards:
ISO 2060: linear density of yarns
ISO 3032, 3933, 1976: measurements of width and length
ISO 5084: measurement of thickness
ISO 3801: measurement of mass
ISO 2062: tensile test for yarns
ISO 5081: strip tensile test
ISO 5032: grab tensile test
ISO 2960: burst test
ISO 4577: resistant against UV-radiation

American standards ASTM:
D 1777-64: measurement of thickness
D 1910-64: measurements of mass
D 3776-84: measurements of mass
D 885: breaking toughness
D 2256: tensile test for yarns
D 1682: strip tensile test
D 2990: creep test
D 3786: burst test
D 1175: abrasion test
D 2261/2262: tear test
D 117 : burst test

Apart from the above-mentioned standards there are British, French and European Standards: BS, AFNOR and CEN.

For impermeable membranes special requirements have been formulated, for which particular test methods have been developed like the puncture test and burst strength test. Besides these 'strength' tests there are tests related to ageing, durability, chemical attack and leaching.

A matting as a ski-loipe.

8.4 *Practical tests*

In some cases it is desirable to perform practical tests. There is a special need for such tests when:
   – Great risks may arise as to the safety of man and environment when the geosynthetic is not succesful in the construction;
   – The project has such a size that for an analysis of costs and profit a detailed specification of the geosynthetic in question is needed;
   – Special requirements are made which cannot be verified with tests or certificates;
   – A reliable general calculation method is not yet available to determine the requirements of the geosynthetic to be applied.
   Practical tests may have various forms. Local circumstances and loading situations have always to be imitated as much as possible to detect the collapsing behaviour. This can be realised by building a test track on the location of the future project or by executing a model experiment in a laboratory at scale 1:1. In a laboratory special attention has to be paid to the imitation of the subsoil.

8.5 *Terms of reference/builders specifications*

In the builders specifications the geosynthetic to be applied has to be described at such a way that the quality of the geo-synthetic is assured, including transport, storage and application method.
   Most of the specifications are addressed to formulate functional demands, criteria and test-procedures after a construction has been realised. But the quality-assurance during execution has also to be described. There is a growing tendency in building specifications to prescribe function demands and less execution prescriptions. In this case the contractor can use his experience and apply the newest technical developments, but the liability has to be settled accordingly. In some countries a standard specification description is in use, which is regularly updated based upon the newest experiences and developments.
   Further the principal might formulate additional conditions regarding the quality assessment of the materials to be applied, the materials to be used and the method of execution.

8.6 *Summary*

To determine the desired quality of a product a package of functional demands will be needed. These demands have to be transferred to material properties of the product concerned, which have to be described in the builders specifications.

To judge if a product meets the requirements samples can be taken and tested. It is also possible to ask for a certificate for the product concerned.

Laboratory and practical tests in situ are performed to demonstrate the suitability of the product and to prove that the geosynthetics meet the requirements.

Practical tests are performed especially when great risks arise in case the geosynthetic fails to be successful.

In the builders specifications the material chosen and the method of quality assessment of both material and application have to be prescribed.

3: Elaborated examples

CHAPTER 9

Embankment on a natural ground with low bearing capacity

9.1 *Description*

For a split-level junction of a proposed route with a railway an embankment is necessary. The height of this embankment is determined by the magnitude of the compression of the sub-base and by the height of the railway arch with which the proposed route has to be connected. With a settlement of 3.5 m the height of the embankment is 11.50 m.

The natural ground consists of a peat layer of 2.50 m with below a 1.50 m peat layer mixed with clay, a 5.00 m layer of clay, a thin layer of clay mixed with peat, a 1.00 m basic-peat and further pleistocene sand.

In Figure 9.1 the soil profile is shown with the embankment before the sub-base has been compressed.

9.2 *Design considerations*

To determine the stability of the embankment the following collapsing possibilities have to be considered:
  – Exceeding of the bearing capacity of the subsoil;
  – Internal collapsing of the embankment in case the slope is chosen at such a degree that the slope may slide over the geosynthetic;
  – Collapsing of the embankment through slip-circle rotation;
  – Collapsing as a result of squeezing.
Not only the stability of the embankment after construction is of importance, but also during construction the stability of the embankment has to be guaranteed.

Stability in the end situation
For the stability in the end situation, when the construction has been finished, the drained strength of the particular sub-base layers is of importance. This is the strength of the sub-base after the water-overpressure has been reduced to normal hydrostatic pressure. In this situation the subsoil has to be built with sufficient

85

resistance, so that the influence of the geosynthetic-reinforcement can be neglected. With respect to the here described example the calculated safety factor is $f_s = 1.35$, according to the method of Bishop, which is sufficient.

See for details Section 4.5.

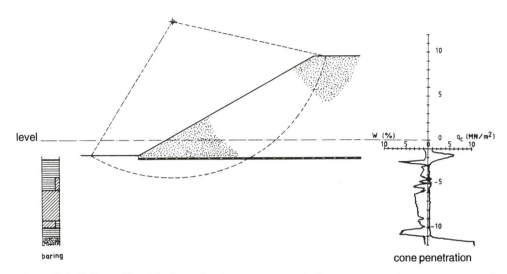

Figure 9.1. Soil profile with the embankment q_c = vertical cone-penetration value, MN/m^2, W = friction number, in percentages of the cone friction value.

Woven bags filled with gravel or sand as a replacement for the railway sleeper-bed.

Stability in the execution phase

During the raising of the embankment the water-pressure in the subsoil increases. The undrained strength of the subsoil is indicative in this case. This is the strength of the subsoil when the overpressured water has not yet been drained. In this phase there is a great danger of loss of stability of the embankment and the subsoil.

The overpressured water can be drained by the raising of the embankment with thin layers, little by little.

If an accelerated execution is desirable or required, a reinforcement with a geosynthetic may give a substantial contribution to the stability of both the construction and the subsoil. The thickness of the layers with which the embankment is built up may then be greater, and consequently the execution phase can be shortened. In combination with an accelerated consolidation by applying vertical drains the realization of the embankment may be quickened.

Because in the example there is a relatively short period in which danger of instability may occur, a safety factor of $f_{s+} = 1.1$ may be chosen. This low safety factor may be chosen in the execution phase, because immediately after the embankment has been finished, the consolidation process begins, resulting in an increase of safety. The use of such a low safety factor requires a careful control, during execution, of water-pressure and deformation measurements. In case the safety factor appears to be too low, the speed of applying new layers has to slow down.

9.3 *Calculations*

From the foregoing it appears that a number of calculations have to be performed to determine the safety. In the following only three situations are elaborated regarding the effect of application of a geosynthetic in an embankment of 8.50 m in height with a slope of 1:2.

This is only one intermediate stage in the construction of the embankment.

By performing of the following calculation, for several layer thicknesses the technically-economically optimum situation can be determined.

The three calculations concern:
– Collapsing according to slip-circle rotation through the sub-soil;
– Internal collapsing by sliding of the embankment over the geosynthetic;
– The loading of the geosynthetic through horizontal deformation of the subsoil by squeezing.

The anchoring length has to be calculated as well.

Collapsing according slip-circle rotation through the subsoil

The calculation of the equilibrium in case of a slip-circle is based on the method of Bishop. Here the safety factor (f_s) is defined as the quotient of the moment of restoring and the moment of disturbing (Fig. 9.2).

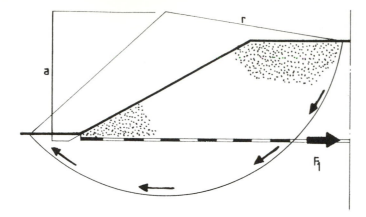

Figure 9.2. Slip-circle rotation in a building phase.

Without application of a geosynthetic reinforcing layer the equation is:

$$f_s = \frac{M_{\text{restoring}}}{M_{\text{disturbing}}}$$

With the application of the geosynthetic the equation is given as:

$$f_s = \frac{M_{\text{restoring}} + \Delta M_{\text{restoring}}}{M_{\text{disturbing}}}$$

$\Delta M_{\text{restoring}}$ is the part delivered by the geosynthetic where:

$$\Delta M_{\text{restoring}} = F_1 \times a$$

where: F_1 = tensile force in the geosynthetic, a = is vertical arm between slip-circle centre and the geosynthetic.

The computer programmes based on the method of Bishop do not contain stability calculations where geosynthetics can be introduced. The influence of a geosynthetic is therefore determined, in this computer programme, as follows:

$$f_{s+} = f_{s-} + \frac{\Delta M_{\text{restoring}}}{M_{\text{disturbing}}}$$

The results of the calculations of the example accordingly to the method of Bishop with a first layer of 8.5, are:
  – The representative slip-circle has a radius of $r = 15.90$ m (Fig. 9.2);
  – The vertical arm a between slip-circle centre and the geosynthetic is 11.20 m (Fig. 9.2).

From soil mechanical calculations can be derived:
- $M_{\text{disturbing}} = 13175$ kNm/m;
- $M_{\text{restoring}} = 12650$ kNm/m;
- The safety factor without the application of a geosynthetic is $f_{s-} = 0.96$.

When a geosynthetic is applied on the original surface

$$f_{s+} = 0.96 + \frac{11.2 \times F_1}{13175}$$

At a $f_{s+} = 1.1$ the tension force in the geosynthetic is derived at as follows:

$$F_1 = (1.1 - 0.96) \times \frac{13175}{11.2} = 165 \text{ kN/m}$$

Internal collapsing through sliding over the geosynthetic

In case a geosynthetic is applied it has to be verified if a sliding over the geosynthetic may occur (Fig. 9.3). For this situation of internal (in)stability it is taken for granted that the safety factor f_s is equal to the quotient of the restoring forces and the disturbing forces.

This means that there must be a equilibrium between the active soil pressure from the embankment (F_{a2}) and the friction over the geosynthetic (F_{p2}):

$$f_s = \frac{F_{p2}}{F_{a2}} = \frac{0.5 \times \rho_{\text{sand}} \times g \times H_t \times \tan \delta \times B_b}{0.5 \times \dfrac{1 - \sin \varphi}{1 + \sin \varphi} \times \rho_{\text{sand}} \times g \times H_t^2} = \frac{B_b \times \tan \delta}{\dfrac{1 - \sin \varphi}{1 + \sin \varphi} \times H_t}$$

Given the properties of the embankment material ($\varphi = 35°$, $\tan \delta = 0.6$ and $\rho_{\text{sand}} = 1750$ kN/m^3) an internal stability with $f_s = 4.4$ can be derived at, which is amply sufficient (embankment height = 8.50 m, slope width $B_b = 17$ m).

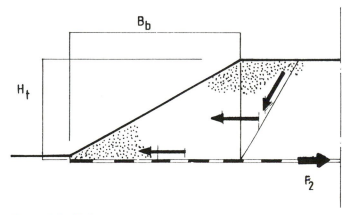

Figure 9.3. Sliding over geosynthetic.

Given the condition that the safety factor has to be larger than 1, the loading of the geosynthetic can never be larger than the loading caused by the active soil pressure. Therefore:

$$F_2 = 0.5 \times \frac{(1 - \sin 35°)}{(1 + \sin 35°)} \times 17.5 \times 8.5^2 = 171 \text{ kN/m}$$

Squeezing
The geosynthetic does not contribute to the prevention of squeezing as discussed in Section 4.5. As a result of horizontal deformation the geosynthetic has been loaded before squeezing arises. Consequently, the geosynthetic cannot be of help against sliding.

To prevent the developing of squeezing a less steep slope may be chosen or a sustaining bank may be constructed.

The loading of the geosynthetic can be estimated with the equation (Fig. 9.4):

$$F_3 = L_k \times C_u$$

Here L_k is the distance between the toe of the slope and the crossing of the sliding circle with the geosynthetic.

For the presented example this results in $F_3 = 93$ kN/m at an undrained shear strength $C_u = 5\text{kN/m}^2$ and at $L_k = 18.5$ m.

Choice of the geosynthetic
In case the geosynthetic has to perform its function during a relatively short period a reduction factor may be used for the influence of creep and mechanical damage.

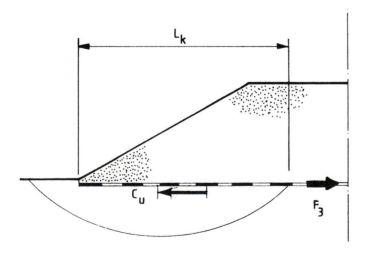

Figure 9.4. Loading of geosynthetic through squeezing.

If the working period of the geosynthetic is determined at 2 years due to creep, a polyester synthetic may be loaded up to 70% of its breaking strength (Table 4.1) and polypropylene up to 20%. If the disadvantageous effect of mechanical damage is determined at 10% the reduction factor (r_i) to determine the required strength of the geosynthetic can be computed for polyester: 1.6 and for polypropylene and polypropylene: 5.5.

The geosynthetic has to be based on the criterium of giving the maximum loading that leads to collapsing. Also the possibilities of the simultaneous arising of phenomena have to be examined.

The calculations of the presented example give the following representative tensile forces in the geosynthetic:
- Slip-circle rotation: $F_1 = 164$ kN/m;
- Internal stability: $F_2 = 171$ kN/m;
- Squeezing: $F_3 = 93$ kN/m;
- $F_{max} = F_2 + F_3 = 264$ kN/m.

The prescribed tensile strength F of the geosynthetic has to be:
- For polyester: $F > 1.6 \times 264$ kN/m ≥ 264 kN/m;
- For polypropylene and polyethylene: $F > 5.5 \times 264$ kN/m ≥ 1452 kN/m.

In respect of the magnitude of the required strength a woven fabric of polyester is advised.

Anchoring length

In the example the geosynthetic is unrolled on the surface. For the calculation of the anchoring length it is estimated that the anchoring arises through friction between the geosynthetic and the embankment. The slight friction between geo-synthetic and the surface is neglected in this case.

The required anchoring length (L_a) can be computed with the following equation:

$$L_a = \frac{F_{max} \times r_i}{\rho \times g \times H_t \times \tan \delta} = \frac{264 \times 1.6}{17.5 \times 8.5 \times \tan 28°} = 5.34 \text{ m (5.5 m)}$$

In tests it has been proved that $\delta = 0.8\varphi = 28°$.

The length of the geosynthetic below the embankment is now:

$$L_k + L_a = 18.5 \text{ m} + 5.5 \text{ m} = 24 \text{ m (Fig. 9.4)}$$

CHAPTER 10

Riverbank protection

10.1 *Description*

To prevent erosion along a river in The Netherlands a quay is built with mine-stone and a sandy sub-base.

Figure 10.1 shows a cross-section of this quay (the measures in m). The representative loading of the revetment is a peak wave height $H_s = 0.65$ m with a wave period $T = 2.4$ s. Based on these figures a quarry stone revetment with grading from 10-60 kg is applied in an amount of 800 kg/m². The layer thickness is 0.5 m.

The other design criteria are:
– Washing out of the subsoil has to be prevented;
– To follow the differences in settlement and some washing-out at the toe of the revetment, the construction has to be flexible.

As a filter a geosynthetic is chosen.

This example deals with the details of the filter construction.

10.2 *Filter-calculation*

Soil-tightness
The sandy subsoil is representative for the required soiltightness. From the average grain-size distribution of some samples (Fig. 10.2) the following characteristic figures are derived: $d_{90} = 330$ μm, $d_{60} = 108$ μm, $d_{15} = 43$ μm, $d_{85} = 240$ μm, $d_{50} = 90$ μm, $d_{10} = 37$ μm.

The internal stability of the soil-structure is determined by:

$$\frac{d_{60}}{d_{10}} = \frac{108}{37} = 2.9$$

As the value is < 10, the soil-structure is stable.

In connection with wind- and ship-waves the hydraulic loading is cyclic.

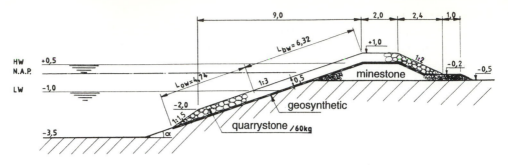

Figure 10.1. Design of a riverbank protection.

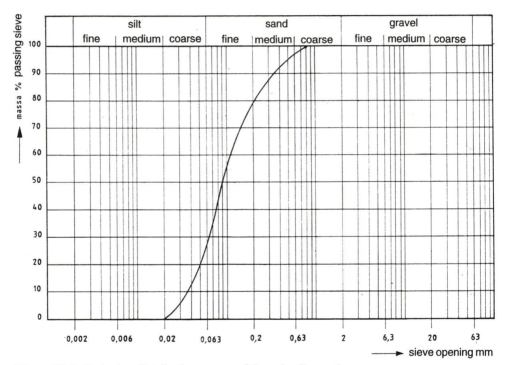

Figure 10.2. Grainsize-distribution curves of the subsoil samples.

Therefore the filter has to fulfil the filter-function criterion:

$$\frac{O(98)}{d_{85}} < 2$$

therefore $O(98) < 480\,\mu m$.

Water-permeability

The water-permeability of the subsoil can be estimated with the data obtained from the grainsize-distribution curves or can be derived at from a test.

The following values can be adhered to:
- For sand: $k = 8 \times 10^{-6}$ m/s;
- For mine-stone: $k = 5 \times 10^{-5}$ m/s.

If one type of geosynthetic is applied it is the mine-stone that is representative in order to determine the permeability of the filter.

As design-criterion is used:
- k(filter) $> 10 \times k$(subsoil) $= 5 \times 10^{-4}$ m/s.

The permittivity ψ of the filter has to be:

$$\psi = \frac{k}{e} = 0.50 \ s^{-1}$$

for a geosynthetic with a thickness of 1 mm

$$\psi = \frac{k}{e} = 0.25 \ s^{-1}$$

for a geosynthetic with a thickness of 2 mm.

10.3 *Strength of the geosynthetic*

The required strength of the geosynthetic has to be derived from an analysis of the collapsing mechanisms both in the end situation after the construction has been finished and during the execution phase.

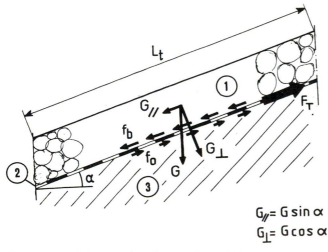

1 Quarry stone
2 Geosynthetic
3 Subsoil

$$G_{//} = G \sin \alpha$$
$$G_{\perp} = G \cos \alpha$$

Figure 10.3. Collapsing mechanism in end situation.

End situation

The representative collapsing mechanism is the sliding down of the geosynthetic together with the quarrystone revetment over the slope.

According to the scheme of Figure 10.3 the following values can be used for the loading of a slip with a width of 1 m:

Above-water table:

$$G = 8.0 \text{ kN/m}^2, L_t = 2.0 \times \sqrt{10} = 6.32 \text{ m}$$

Under-water table:

$$G = 4.9 \text{ kN/m}^2, L_t = 1.5 \times \sqrt{10} = 4.74 \text{ m}$$
$$\tan \alpha = 0.33$$
$$f_o = 0.3 \text{ (geosynthetic-subsoil)}, f_o \text{ is chosen so low for the sake of the example}$$
$$\begin{aligned} F_T &= (8.0 \sin \alpha - 8.0 \cos \alpha \times f_o) \times 6.32 + (4.9 \sin \alpha - 4.9 \cos \alpha \times f_o) \times 4.74 \\ &= (8.0 \times 6.32 + 4.9 \times 4.74) \times (\sin \alpha - f_o \cos \alpha) = 73.79 \times 0.032 \\ &= 2.36 \text{ kN/m} \end{aligned}$$

The most unfavourable situation arises when $f_o = 0$;

$$F_T = 73.79 \sin \alpha = 23.33 \text{ kN/m}.$$

Woven bags filled with soil, sand or gravel to construct steep slopes, bank revetment or current guidance.

Because of the lasting functioning of the filter construction and due to the less favourable creep characteristics of polypropylene, the application of a polyester woven fabric is preferable.

As indicated in Section 4.2 a reduction factor $r_i = 2.3$ has to be used to determine the required strength of the geosynthetic.

Consequently, the required strength of the geotextile in the most unfavourable situation is:

$$F_{max} = 2.3 \times 23.33 = 53.7 \text{ kN/m}$$

Execution phase

The loading of the geosynthetic during construction mainly depends on the working method used. Concentrated tensile forces, friction and punching have to be prevented.

In dry situations the geosynthetics are unrolled so that additional loadings may be limited.

In wet situations (applications below the water table) mattresses may be loaded unequally during transport from the building site of the mattress to the point of sinking. This unequal loading may in part be prevented by use of a stiff frame (Fig. 7.1).

For bank protection geosynthetics are applied with a mass $> 0.2 \text{ kg/m}^2$ and a tensile strength $> 15 \text{ kN/m}$.

10.4 *Remaining aspects*

Heat development at the building site

A mattress is mostly built at a separate building site. When such a mattress is thrown off, heat may develop through friction. This heat may melt the geosynthetic. To prevent this the building site may be watered or covered with steel plates to reduce the friction and to convey the heat.

Gliding of ballast

The risk of gliding of ballast (quarry stone or riprap) over the geosynthetic is always present if the friction-coefficient f_b between geosynthetic and ballast is lower than tan α. In most cases also f_o is critical, owing to the possible risk of sliding of the complete revetment over the sub-base. This can be prevented by anchoring of the geosynthetic on the upper side of the slope.

In the presented example the gliding of ballast is not critical ($f_b > \tan \alpha$). In the cross-section (Fig. 10.1) an additional security is obtained by covering the whole mine-stone quay with the geosynthetic.

10.5 *Choice of the geosynthetic*

From the above mentioned phenomena it appears that besides the main functional requirements as a filter layer, soil tightness and permittivity, the requirements for strength also determine the kind of geosynthetic to be applied. In some cases it is also advisable to make demands on the durability of the geosynthetic to be prepared for UV-radiation and thermic oxidation.

The main requirements can be summarized as follows:

– Soil-tightness: $O(98) > 480\,\mu m$;
– Permittivity: $\psi > 0.3\text{-}0.5\ s^{-1}$, depending on e;
– Tensile strength: $F > 54\,kN/m$.

With the help of product information sheets a choice can be made from the most suitable geosynthetics.

CHAPTER 11

Landfill

11.1 *Description*

For a depot of chemical refuse a landfill has to be constructed in such a way that the refuse is isolated and the effectiveness of isolating can be controlled and governed.

The subsoil of the chosen location consists of sand and the groundwater table is at 7.50 m below surface.

The upper side of the landfill has to be finished at 15 m above the existing surface. Moreover, the area is susceptable to settlements.

11.2 *Design*

The function of the depot is to storage the chemical refuse in such a way that the risks of emission of toxic materials is reduced to acceptable limits. In fact, the design and the execution should be based upon a risk analysis. This kind of design is not yet in use generally, and therefore the design of landfills for refuse is based on local regulations which differ from country to country.

The most important element in the construction of a landfill is the sealing at the top-side, at the side and at the bottom. There may be no transport of water from the surroundings (rain- and groundwater) to the refuse and no drain of water and chemicals from the refuse to the environment.

In respect of the high requirements made on the durability of the landfill, additional constructive measurements have to be taken to minimalize the fail chance. For instance, due to the weight of the landfill, unequal settlements may occur if the subground is inhomogeneous, causing overstressing of the geosynthetic and consequently damaging.

Therefore, in most situations a watertight geosynthetic is combined with a tight mineral layer (bentonite).

To prevent overpressure through gasses arising in a landfill of mixed waste, a degassing layer is constructed; and to control and eventually to drain possible leakage a layer is applied on the bottom of the landfill. To assure a reliable execution and to make a gradual transition from coarse refuse to the geosynthetic, the bottom layer is covered with fine- graded material. To prevent penetration of groundwater the bottom of a landfill has to be laid at least 0.70 m above the highest possible groundwater table.

In the example a difference of 1.00 m is chosen so that the bottom of the landfill lies at a depth of 6.50 m below the surface. To gain a landfill with the largest possible contents the slope in the cutting is chosen at 1:1½. Above the surface the slopes of the landfill are chosen at 1:3 due to reasons of landscape and execution.

As the subsoil is sensitive to settlement the construction is always riskful.

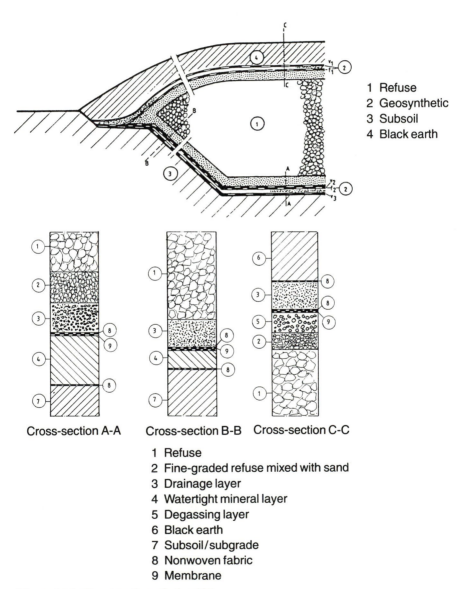

1 Refuse
2 Geosynthetic
3 Subsoil
4 Black earth

Cross-section A-A   Cross-section B-B   Cross-section C-C

1 Refuse
2 Fine-graded refuse mixed with sand
3 Drainage layer
4 Watertight mineral layer
5 Degassing layer
6 Black earth
7 Subsoil/subgrade
8 Nonwoven fabric
9 Membrane

Figure 11.1. Construction of a landfill.

Sealing of the bottom of a landfill

Owing to a groundwater level at 7.50 m below the surface the cutting can be made without draining.

Because the landfill is situated in a riskful location, the bottom has to be provided with a mineral tightening layer combined with a membrane. The construction of the bottom is now as in Figure 11.1, cross-section A-A.

The mineral tightening layer consists of a mixture of sand and bentonite with an additive to increase the resistance against chemical attack, for instance water-glass.

By adding this additive the k-value of the mineral tightening layer decreases from $k = 10^{-9}$ m/s to $k = 10^{-11}$ m/s.

The membrane to be applied has to be a HDPE membrane with at least a thickness of 2 mm. In many countries additional requirements are made regarding the thickness and the strength of the membrane.

To prevent damage as a result of coarse drainage material the membrane may be covered by a thick nonwoven fabric, voluminous and chemically resistant. The level of the percolation water in the drainage-layer has to be held so low that the water table can not exceed above the drainage layer with a thickness of 0.50 m.

This can be realized by applying a chemically resistant dewatering system in the drainage layer. For this purpose pipes are placed in the drainage layer connected with a pump outside the landfill. The diameter of these pipes have to be large enough to admit inspections by TV-cameras and for maintenance as cleaning of the drain pipes by water pressure.

Owing to the weight of the landfill the drain pipes must be able to resist a considerable external pressure.

Anchoring of a membrane at the topside of a landfill in cutting.

Separation layer between sub-base and black earth for overgrowth in a tram-track.

An early application of a geotextile in the Netherlands (Pluimpot, 1953): nylon bags as support construction for a crane-track and pressure pipe (1950).

Slope in cutting

Because the slope in cutting is chosen at 1:1½ it is necessary to reinforce the tightening layer to prevent the arising of large tensile forces in the membrane. The strength of such a reinforcement can be calculated as indicated in 10.3. The slope construction is now as in Figure 11.1, cross-section B-B. The drainage layer cannot be applied beforehand, because this layer will slide on a slope of 1:1½.

This drainage layer has to be applied together with the filling of the landfill.

Top sealing

As soon as the landfill reaches the level as foreseen, the covering layer is placed (see Fig. 11.1 cross-section C-C).

Owing to the flat slope of 1:3 there is no need of a reinforcing layer. The construction of the top sealing and that of the slope may be identical. Due to the expected deformation of the slope a 2 mm thick HDPE-membrane is applied. This membrane is provided with a printing to increase the friction resistance.

In landfills for mixed domestic-waste gasses develop through activities of micro-organisms. These gasses have to be transmitted and therefore a degassing layer is applied consisting of coarse broken stone.

11.3 *Fixing and jointing*

To discharge percolation water, leachate and gas venting and for inspection pits the membrane has to be perforated. These perforations have to be made in such a way that these spots can be controlled continuously. Of course, the amount of perforation has to be restricted as much as possible. The membrane at the bottom has to be provided with slabs on which the membrane of the slope and/or topsealing can be welded.

Literature

1. Reference books:
   - NGO, *Geotextiles and Geomembranes in Civil Engineering,* A.A. Balkema, Rotterdam, 1995.
   - Rankilor, P.R., *UTF Geosynthetic Manual,* UCO Technical Fabrics, Lokeren, Belgium.
2. *Vergelijkende Produktencatalogus Geotextielen* (Product catalogue of geosynthetics), Nederlandse Geotextielorganisatie, 1994.
3. Stichting C.R.O.W., Publikatie 27, 1989, *Geotextiel in de aardebaan* (Geotextiles in embankments).
4. Sellmeijer, J.B., C.J. Kenter & C. van der Berg, *Calculation method for fabric reinforced road,* S.I.C.G., Las Vegas, 1982.
5. Stichting C.R.O.W., Publikatie 28, 1989, *Geotextiel onder wegfunderingen* (Geotextiles in road foundations).
6. Brinch Hansen, J., *A General Formula for Bearing Capacity,* Ingenioren-International Edition, Geoteknisk Institut Bulletin Nr.11.
7. Jewell, R.A., *Application on the Revised Design Charts for Steep Reinforced Slopes,* Report No. QUEL 1796/89, University of Oxford, January 1990.
8. *Handboek oeverbeschermingsconstructies* (Handbook riverbank revetments), Nederlandse Vereniging Kust- en Oeverwerken, Rotterdam, 1986.
9. *Kunststoffilters in Kust- en Oeverwerken* (Geosynthetic filters), Nederlandse Vereniging Kust- en Oeverwerken, Rotterdam, 1986.
10. Stichting C.R.O.W., Technisch rapport, 1991, *Geotextielen, eigenschappen en testmethoden* (Geotextiles, Properties and Standards).
11. Stichting C.R.O.W., *Standaard 1985 RAW Bepalingen* (Building specifications).
12. Hoeks, J., H.P. Oosterom & D. Boels, *Richtlijnen voor ontwerp en constructie van eindafdekkingen van afval- en reststofbergingen* (Guidelines for design, Construction and Sealing of Landfills), Staring Centrum, Wageningen, 1990, Rapport nr.91.
13. Stichting CUR, publikatie 174, *Geotextielen in de waterbouw* (Geotextiles in hydraulic engineering), CUR, Gouda, 1995.
14. Stichting CUR, publikatie 175, *Geokunststoffen in de wegenbouw en als grondwapening* (Geosynthetics in road construction and as soil reinforcement), CUR, Gouda, 1995.
15. Stichting CUR, publikatie 176, *Geotextielen in afval- en reststofbergingen* (Geotextiles in waste and residue storage), CUR, Gouda, 1995.

6-27-96

ENGINEERING